URBAN
AMERICA IN
TRANSFORMATION

URBAN AMERICA IN TRANSFORMATION

Perspectives
on Urban Policy
and Development

BENJAMIN KLEINBERG

SAGE Publications
International Educational and Professional Publisher
Thousand Oaks   London   New Delhi

For information address:

SAGE Publications, Inc.
2455 Teller Road
Thousand Oaks, California 91320

SAGE Publications Ltd.
6 Bonhill Street
London EC2A 4PU
United Kingdom

SAGE Publications India Pvt. Ltd.
M-32 Market
Greater Kailash I
New Delhi 110048 India

Printed in the United States of America

Library of Congress Cataloging-in-Publication Data

Kleinberg, Benjamin.
    Urban America in transformation: perspectives on urban policy and development / Benjamin Kleinberg.
        p.   cm.
    Includes bibliographical references and index.
    ISBN 0-8039-5295-3 (cl).—ISBN 0-8039-5296-1 (pb)
        1. Urbanization—United States. 2. Urban policy—United States.
    3. Cities and towns—United States—Growth. 4. City planning—
    United States. I. Title.
    HT384.U5K56   1995
    307.76'0973—dc20   94-3429
                                                                   CIP

95 96 97 98 99 10 9 8 7 6 5 4 3 2 1

Sage Production Editor: Yvonne Könneker

Contents

Part II: Evolution and Organization of the Urban Policy System

Preface

This book is a response to my discontent with two major perspectives on postindustrial urban policy and development: the contemporary (or *neo*) ecological perspective and the neomarxist perspective. Each seeks to explain the underlying dynamics of the emerging new urbanism of the post-World War II era in America, one as a space-economy of urban and suburban development and the other as a political economy of those developments. Yet although each makes reference to other dimensions of analysis, both are primarily economic theories of development. Despite the significance of public organizations, policies, and programs for urban development, neither perspective provides a sufficiently articulated analysis of the intergovernmental urban policy system as a framework for the locational issues, class conflicts, or systemic imperatives that drive their respective theories.

The first part of this book begins with critical analyses of those approaches and ends by introducing a third, interorganizational/policy perspective. It is presented not as an alternative theory but as a structural framework for analyzing the shaping and implementation of urban policy. In the absence of such a complementing framework, I believe that neither of the other perspectives fully can analyze or explain the phenomena that each addresses. The basic dimensions of interorganizational/policy analysis can be stated in terms of the *horizontal* (local-level) and *vertical* (federal-state-local) interorganizational relations of the contemporary urban policy system. The essential point is that unlike the earlier 20th century, such an analysis today is incomplete unless it conceptualizes and examines the role of the vertical dimension, which connects the development of urban localities to federal or state policies and programs of significance to contemporary urban trends and needs.

The second part of the book is devoted to a thorough examination of the emergence and development of a vertically organized urban policy system and its changing suitability over different policy periods for delivering programs targeted, for example, to economic development versus programs targeted to social renewal, especially in the social/spatial restructuring of formerly industrial big cities. Chapters 4 through 6 trace the development of the federal urban policy system from its origins in the rise of New Deal *Cooperative Federalism*, to its operation in the urban renewal program, and then to a modified ("creative") form of Cooperative Federalism in the War on Poverty. Chapter 7 analyzes the early 1970s' reorganization of the federal dimension of urban policy under the Nixon administration's *New Federalism*, developed as a major policy and organizational reaction to the social policy objectives and implementing program structures of the War on Poverty. This is followed by discussion of the Carter administration's attempt to create a *New Partnership* (one both intergovernmental and public-private) to serve a shifting mix of antipoverty and economic development objectives.

The book concludes in Chapter 8 with an analysis of the Reagan/Bush version of New Federalism, which aimed at producing a fundamental restructuring of urban and related social policy through a combination of program reorganization and fiscal strategies including sharp budgetary retrenchments. It examines the responses of states and cities to this new, more constricted variety of federalism, including the resort by many cities to local public-private partnerships, which are analyzed in terms of a typology ranging from business-dominated partnerships to partnerships structured on the basis of more equitable strategies for planning local development.

This book is intended as a text for upper-division undergraduates and master's level graduate students with interests in urban sociology, urban politics, urban planning, public administration, community-oriented social work, and intergovernmental relations. It is aimed also at public officials, administrators, and staff and at community activists and lobbyists for urban needs. Finally, I hope to direct the attention of at least some part of the educated lay public to the issues addressed, so that a developing movement for progressive urban change may find some response in the broader public.

For all these audiences, the central aim of the book is to stimulate thinking in terms of how the organization and objectives of federal programs and local public-private partnerships make a difference; how they reflect the incorporation or exclusion of different interests, values, and goals for urban development; and how they have influenced public policies that significantly affect the social-spatial patterns of city life in America. Urban development is not a mechanical process determined by forces of technological innovation or by inevitabilities of class conflict; it can be moved in socially positive directions by concerted public action. In short, assisted by an understanding of the urban development process and its organization, concerned citizens and their advocates can influence public policy to pursue social goals and community values in the further development and renewal of American cities.

ACKNOWLEDGMENTS

Here I must express my appreciation to several colleagues who have read sections of the book and made helpful comments: Howell Baum (Urban Studies and Planning Program, University of Maryland, College Park), Jay Sokolovsky, and Derek Gill, who as my department chairman generously has nurtured this project with his continued interest and support. In addition, I wish to express my thanks to Richard Ullrich, chair of the economic policy committee of the Baltimore Development Commission, for enabling me to participate in several informative meetings with members of the Baltimore Planning Department, as well as meetings with Mayor Schmoke concerning the commission's support for a Baltimore Works proposal that addressed a number of issues discussed in this book. Many thanks are due also to Carrie Mullen at Sage, who despite fires, floods, and earthquakes has been a remarkably steady and helpful source of editorial support. Finally, my greatest debt of gratitude is to my wife, Susan, for her continuing assistance and encouragement in what may at times have seemed a project without end. Thanks to her, its demands managed to bring us together more than they kept us apart. And to Allen, Leah, and Kim, thanks for being the generation of hope that makes this all worthwhile.

Introduction

During the past 50 years the urban landscape of America has been rebuilt. Massive suburbanization on the one hand and downtown renewal on the other wholly have changed the shape of American urbanism. From a nation of a limited number of large industrial cities and a multitude of small rural towns, the United States has become a society of metropolitan areas. These areas consist of central cities surrounded by suburban zones that now include not only sprawling residential suburbs and older small towns, but a growing number of full-scale suburban cities as well. The specific "ecological patterns" characteristic of American urban populations—the distinctive manner in which they use land in cities and their surrounding suburbs for residence, commerce, industry, transportation, and recreation—have changed fundamentally from the compact, densely populated cities of the early 20th century, then located mainly in the Northeast. Today's postindustrial urbanism is represented by a wide-ranging, decentralized metropolitan form of settlement, which in scale dwarfs the earlier concentrated urban-industrial settlements and in extent now stretches from coast to coast, from north to south.

This is the visible face of urban change. But behind the scenes, underlying these great physical and social changes there has been another kind of restructuring. That is the restructuring of the major economic and political organizations of American society and their institutional interrelations, all of which have provided the organizational context for contemporary urban and suburban development. In short, the spatial restructuring of urban America cannot be understood simply by references to the construction of physical facilities such as downtown office buildings or suburban shopping malls or to the movement of populations from cities to suburbs or from region to region, which are so much stressed in contemporary ecological theory. Urban change cannot be explained solely in terms of the action of economic or related technological factors such as the evolution of high-speed highways and communication networks, however significant their contributions have been to the decentralization of once-compact forms of urban settlement. As I will discuss, overemphasis on economically related factors has left relatively unexplored the role played by government in setting the framework for contemporary urban change, not only in the mainstream urban ecological perspective, but also in the marxist urban studies that have provided the major critique of that perspective.

The story behind the physical restructuring of urban America is to a considerable degree one of changes in the organizational structure and operation of America's intergovernmental system. The central focus of this book is its description and analysis of the federal system of urban-related policy making as an evolving network of interorganizational relationships. This book reviews the development of this system since the 1930s and explores its significance, particularly for the postindustrial transition of older big cities.

The book begins with a critical review of the established ecological and neomarxist perspectives on urban development. It goes on to provide a policy-oriented conceptual framework as a context within which the developments they describe can be understood more fully. A key theme is that the urban policy system is a politically sensitive intergovernmental network for administering the delivery of urban programs to the local level. Its changing organizational structures and program goals reflect the shifting influence of competing policy interests in American society. Depending on how the urban policy system becomes structured and what interests it incorporates, it will encourage certain types of urban policy while blocking others during extended periods of policy dominance.

Applying this perspective, the book first explores the evolution of policy objectives and program organization in relation to the urban policy system that originated in the context of New Deal Cooperative Federalism. It traces the impacts of programs created under the influence of Cooperative Federalism, beginning with public works and housing, on through programs for central city business district renewal, and then to the programs of the War on Poverty. It then examines the move to a New Federalism first established by the Nixon administration and further developed during the Reagan era. A major issue explored here is how the policy objectives and structures associated with the New Federalism have played a critical role in enabling the shift from programs focused on inner-city concerns to a more diffusely suburban/metropolitan orientation for community development.

Throughout, this book emphasizes that the choice of federal policies (e.g., downtown corporate center redevelopment versus neighborhood revitalization) and the structuring of programs that implement those policies each can have important consequences for urban development, affecting not only changes in land use or construction of new facilities, but also employment opportunities and social status of populations in central cities and their suburbs. Not least, it points out that how public programs are organized intergovernmentally affects both the relative political access and distribution of program benefits to competing groups in the policy system, influencing both the underlying process and the visible products of urban policy.

At this point, many questions exist as to the fate of urban-relevant policy for the foreseeable future. In the retrenchment phase initiated during the Reagan era, urban policy turned away significantly from social program objectives and their traditional intergovernmental channels. Has this meant the end of the federally dependent intergovernmental city or, instead, perhaps its transition to a new set of dependencies? Is it possible under conditions of stringent federal budgetary constraint to do justice to either the renewal of urban economic infrastructure or the provision of needed social programs? Can a balanced and effective urban policy agenda yet be established, incorporating initiatives for both economic development and social upgrading? And what respective roles might states and localities now play, in relation to reduced federal program efforts on one hand and to increasingly mobile urban business on the other? These are large questions, which cannot all be fully answered in a single work. However, posed within a suitable frame of reference, they may at least be better understood. This book will examine these questions with an eye to future prospects for urban policy in older large cities, taking into account the impact of federal retrenchment policies of the past decade.

PART ONE

Urban Development in Perspective

[1]

The Ecological Perspective

THE RISE OF THE INDUSTRIAL CITY

In their relatively brief history, American cities have experienced several important transformations. The first major transformation ushered in the development of large industrial cities, the very type of cities that have been the focus of discussion of urban crisis in contemporary times. This period began roughly during the 1840s, about a half century after the Republic had been established. Within the next half century the distinctive features of the early industrial city were developed.

The Preindustrial Background

Prior to the 1840s the only sizable American cities were the port cities of the East Coast. Cities such as Philadelphia, New York, Boston, and Charleston ranked as the leading urban places at the country's national beginnings in the late 1700s. These were actually small commercial cities under 30,000 in population, built up during the colonial era around seagoing ports to serve the mercantile policies of the British empire. Chartered by the British Crown, their major purpose was to control and channel the trade between Old and New Worlds, placing the profits that arose from it in the hands of a British commercial elite. These urban places were therefore of some importance as colonial trading centers in an imperial network of controlled commerce. Together with about a dozen other towns and small cities on the East Coast, they operated as market centers for their neighboring rural hinterlands and in some cases as military garrisons or administrative headquarters for the colonial government (Glaab & Brown, 1967).

The typical preindustrial city or town was a modestly scaled "walking city" in which transportation for the most part was by foot and housing rarely rose more than a few stories. It was a settlement usually no more than a mile in radius, hence small enough to be bound together by personal networks of family, friendship, or business relations. Accordingly, social organization typically was based on primary relations involving frequent personal communication and face-to-face contact between residents. Indeed, not only everyday social life, but politics as well, was organized around kinship and personal clique relations with the early towns being governed by members of the "better families," who not merely were wealthy and prestigious but also involved themselves directly in the management of public affairs (Goldfield & Brownell, 1979). Finally, with town life based on family status and relationship, local governments provided only the most basic public services, and the public sector, what there was of it, was not yet differentiated clearly from private life. The few elementary public services such as sanitation, fire fighting, and police protection generally were provided through cooperation between individual families who joined together in private voluntary organizations often built around friendship or neighborly relations (Glaab & Brown, 1967).

The Urban-Industrial Transformation

The great historical transformation from preindustrial towns barely large enough to be considered cities to full-sized, even giant industrial urban centers was a notably rapid one. By the mid-1800s developments in transportation, such as the growth of steamship links between American and European ports and the building of an extensive network of railroads in the United States, made possible the massive European immigration that came in response to expanding opportunities for employment or land in America, particularly after the Civil War. In turn, these events reflect the successful development of American capitalism from a predominantly commercial to increasingly industrial form. This evolution was manifested in the growing volume of trade passing through East Coast seaport cities such as New York and Philadelphia after Independence, followed by the investment of increasing sums in factories and railroads throughout the urbanizing industrial belt from the Northeast to the Midwest (Ward, 1971). Beginning with the expansion of overseas trade after Independence and the wide investment of the profits in transportation and industrial networks, this dynamic process soon led to the rise of numerous urban centers tied to outlying rural markets by a combination of roads, canals, and rails. Further intensified investment in industry and railroads spurred the growth of a large-scale factory system and made for a fast-paced development to the emerging system of industrial urbanism that came into its own by the late 19th century.

The speed of this transformation is evidenced by the fact that in over 200 years of settlement prior to 1840, America's urban population never rose much higher than 10% of the total population, while in the next 60 years it grew to almost 40% of the total, clearly challenging any lingering Jeffersonian notions of small town dominance. In 1840 there were only three cities over 100,000 population in the entire country. During the next two decades the number of such cities more than doubled, with one (New York) going beyond 1 million. By the onset of the Civil War in 1860, the urban percentage of the nation's population had grown to nearly 20%, an increase of about 9% in the two decades from 1840 to 1860, compared with less than 6% over five decades from 1790 to 1840 (U.S. Bureau of the Census, 1957; U.S. Census of Population, 1970).

In the aftermath of the Civil War, rapid urbanization became a visible fact of life, so that many have referred to the war as a watershed both in American urbanization and in broader social history. Like no other arena, the war provided a historic context for direct and intense confrontation between the urbanizing, industrially expanding North and the agrarian plantation South as alternative socioeconomic systems. The victory of the North meant a clear path for laissez-faire capitalism on a national scale and ensured the continuation of rapid industrialization and of urban development in the vicinity of industrial centers.

Transportation and Urban Development

As noted earlier, the rapid growth of cities, particularly middle- and large-sized cities, and the expansion of urban population after the early decades of the 19th century can be traced to several factors that were significant to the development of industrial capitalism in America. First was the transportation revolution that began with the laying down of a railroad network linking the industrializing cities of the North to each other and to the resource-rich hinterlands of the South and Midwest. By the beginning of the Civil War, a network of over 30,000 miles of railroad had been set in place, tying together the urban nodes of the major populated regions of the country and poised to begin moving into the Far West, there to complete the outlines of a continental rail system by the end of the 19th century (Ward, 1971).

From the perspective of economic development, the construction of the railroads was a decisive step forward in linking together the basic factors of economic production, at first on a regional and then on a national scale. Through the construction of a far-flung rail network with urban communities located along its lines and at major transport hubs such as Chicago, Cleveland, and Kansas City, it became possible to penetrate the interior of the continent and make available vast resources of land, raw materials, labor, new technologies, and the financial capital by which these might be acquired and brought together. In sum, the installation of the railroads made it possible for all the major factors of production to be brought within convenient mutual access to one another and for their products to be set circulating into ever widening markets in a rapidly developing urban-industrial system.

Immigration and Urban Development

Another major element in the urban development of the industrializing United States was the accelerating flow of immigrants from the European continent. This was to have important consequences not only for the industrialization process but also for various aspects of later urban social development.

The bulk of the European immigration took place during the course of the century from 1830 to the Depression of the 1930s. This great immigration occurred in two distinguishable waves, later referred to as the "old migration" (approximately 1830 to 1880) and the "new migration" (1880 to 1930). The most important sources for the first wave were the countries of western Europe, particularly England, Ireland, and Germany, as well as the countries of Scandinavia, which became significant sources of immigration after the Civil War. These western European countries continued to be the main points of origin for the European immigration until the 1880s. During that decade, the major source areas began shifting to southern and eastern Europe, particularly Italy, the Balkans, Russia, and Poland. By the early 20th century these more remote sections of Europe had become the predominant sources of a massive immigration without historical precedent, overall tripling the average annual levels of influx of the old migration.[1]

Whether escaping from crop failures and famines or from political repression and religious persecution, the massive emigration of these "poor, huddled masses" served as a major source of labor supply for the dynamically expanding industrial capitalism of the United States, which, unlike the European states, was not hampered by permanent barriers of language differences or conflicts between different nationalities from spreading across the continent to fulfill its "manifest destiny" as the leading industrial country of the world. Once the Civil War had settled the issue of national union, the linked processes of large-scale industrialization and urbanization were free to move across the country, and they did so quite rapidly.

The Industrial City

The industrial city that emerged from these developments was massive in size, densely settled, and highly mixed in regard to the populations that it included. It was a far more complex and fast-moving urban setting than the typical preindustrial city. That city had been regarded as a place of amenity and culture, as well as the seat of government and commerce. In the industrial city this view underwent a radical change. Now the city, particularly its central business and industrial district, came to be regarded as a great economic machine, a meeting point for the factors of production—land, labor, raw materials, and business organization—from whose successful combination the investor or entrepreneur could turn a handsome profit.

From the outset, the developing industrial city became the center of private investment in a growing factory system and of private and public investment in basic support facilities (infrastructure) for industry as well as for residential growth and the daily conduct of urban life. Infrastructure facilities ranged from streets, roads, and public lighting to water and sewage systems, transportation, and energy utilities, and the physical plant and equipment required by activities such as public education, police services, and fire protection. These physical facilities and the services that they supported were basic to the development and functioning of the industrial city and the health, convenience, and skills of its population. Investment in these facilities drew vast amounts of domestic and even foreign capital; in turn, this contributed not only to the cities' expansion but also to the expansion of the national economy because the new urban centers were mass markets for consumer goods and services as well as for capital or production goods; for the products of the farm and the services of commerce as well as for the output of industry (Glaab & Brown, 1967; Kirkland, 1970).

Historically, one of the economic functions of the city has been its role as a central marketplace. The populations of preindustrial cities and towns provided markets for the purchase of food and raw materials from the countryside, as well as for the handicraft products of their own resident craftspeople. Later the industrializing city added markets for labor, for manufactured goods, and for capital, the last typically in the form of loans and credit with which the other factors of production could be purchased. These market functions often were expanded significantly as cities grew in late-19th-century America. Thus, to the extent that the city of the industrial era was also a complex market and commercial center, it was not simply an industrial city.

After the Civil War, growing industrial cities tended to develop quite diversified industrial-commercial economies, including not only many types of manufacturing but also various types of commercial activity and the provision of related personal and business services. The large cities of this era evolved increasingly into industrial-commercial centers, functioning both as key locations of factory production and as multipurpose marketplaces through which their growing populations could access locally produced consumer goods and services and their growing industries could purchase raw materials, equipment, and the services of broad pools of labor. The larger the industrial era city, the more diverse its functions and the larger and more differentiated the workforce of its industries and businesses. Drawn by the opportunities for employment offered by an expanding factory system, European immigrants flowed into America's industrial cities for over a half century from the end of the Civil War to the Great Depression. The result was a great diversity of populations, cultures, and lifestyles over and above the various occupations and classes that might be expected to be found in industrial centers.

In short, the industrial city was a highly complex urban settlement, involving a multitude of different, often competing economic and social interests. How to comprehend all of this complexity was one of the major tasks facing urban social theorists of the early 20th century. The result of their efforts is known today as classic ecological theory. This theory introduced a perspective that remained central to urban sociology until at least the 1950s and that is influential in urban planning even to the present day.

CLASSIC URBAN ECOLOGY

In the 1920s University of Chicago sociologists Robert Park and Ernest Burgess introduced the concept of *human ecology* to American social science (Park & Burgess, 1925). The field of ecology had originated in the 19th century as the branch of biological science concerned with the adaptation of plants and animals to each other and to their natural surroundings. This provided the distinctive viewpoint of classic urban ecology, which therefore can be traced to the English biologist Charles Darwin's conception of an organic *web of life*. In that web, different natural species relate to each other in a complex network of competition and interdependence, set within the specific context of their local environment. Within its environment, each species develops a particular territorial position, an ecological niche from which it seeks to ensure its survival against competing organisms relative to the resources available in the environment. It was the Chicago sociologists who applied this organismic perspective to the human community and specifically to the urban community.

A central theme of the classic theory was that without anyone planning it, the new urban order tended to evolve spontaneously on the basis of competitive processes similar to those that can be identified in the struggle for survival in nature. This analogy seemed quite appropriate because the industrial city was in fact the scene of pervasive competition among both immigrant and established groups for land and other resources. Thus an image of urban evolution based on continual social and spatial competition resembling that found among natural organisms provided the roots of classic urban ecology.

Studying the development of the industrial city, Park and Burgess (1925) observed a struggle between competing populations and interest groups for the points of best advantage with relation to the valued resources of the environment. At stake were physical and social access to desirable housing and places of employment as well as to commercial and public facilities important to the well-being and advancement of group members. In this division of urban resources some groups succeeded in achieving a position of ecological dominance. Park (1952) points out that in the plant community the dominant species has the special function of imposing ecological order over the numbers and distribution of all the various types of plants in that community. Similarly, in the human community "the principle of dominance ... tends to determine the general ecological pattern of the city and the functional relation of each of the different areas of the city to all others" (p. 152).

Adapted by simple analogy with the natural environment, concepts such as *ecological order* and *dominance* became crucial to the perspective of human ecology. In particular, *dominance* is that condition in which some local competitor acquires a strategic location from which it exerts controlling influence over the surrounding territory, establishing what amounts to a local hierarchy in land use. In the modern city, the central area of dominance typically is the area of highest land values. Dominance in this area is available only to a limited number of competitors in the urban system, specifically those who can afford the high land costs associated with such a point of advantage and also can make the most economically efficient or profitable use of the location. In most large cities, the central business district is such an area, and the dominant land uses are associated typically with major banks, corporate office buildings, department stores, theaters, and other leading business, commercial, or entertainment uses.

It is clear that dominance is a phenomenon that has important consequences for the whole ecological system. In the urban community, Park (1952) saw dominance arising from the ongoing "competition of life itself," placing each urban actor "into the particular niche where it will meet the least competition and contribute most to the life of the community" (p. 161). In this view, from the "natural" and unregulated process of competition for physical and social position in the community there emerges spontaneously an organically ordered social-spatial system in which each participating group makes its best contribution, so that the spatial positioning of different groups turns out to be functional for the good of the community as a whole.

The Functional Order of the City: Theory of Concentric Zones

It was in the work of the University of Chicago urban sociologists of the 1920s (the Chicago school) that the image of the functional order of the industrial city was expressed most clearly. According to these classic sociologists, the industrial city could be understood as a series of concentric zones of different land uses surrounding the *central business district* (CBD) (Park & Burgess, 1925, pp. 54-56). The CBD was seen as not only the physical core but also the functional heart of the city. In fact, the basic structure and dynamics of the city's development were viewed as deriving from the continuous expansion of the CBD into surrounding areas (see Figure 1.1). Hawley (1971) summarized this conception as follows:

> Growth of the central business district pushes ahead of it a belt of obsolescence occupied by light industries, warehouses, and slums. This transition zone, in turn, encroaches on a zone of low-income housing, causing the latter to shift outward and to invade a belt of middle-income residential properties. . . . The occupants of each inner zone tend to succeed to the space occupied by those of the next outer zone. At any moment in time, therefore, the distribution of land uses exhibits a ring-like appearance. (p. 99)

The encroachment of occupants and facilities from one zone on the next was referred to by the Chicago ecologists as *invasion*. The culmination of this process with the replacement of former residents by newcomers and of one set of land uses by another was known as *succession*. Yet, although invasion and succession involve movement from one zone to another, generally they

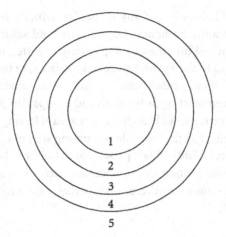

1. Central   2. Zone in   3. Working-class 4. Middle-class   5. Suburban
   Business and   Transition   Residential   Residential   Commuter
   administrative   (from   Zone   Zone   Zone
   District (CBD)   residential to
                     nonresidential)

Figure 1.1. The Classic Chicago School Concentric Zone Model

SOURCE: Based on "The Growth of the City," by E. W. Burgess, in R. E. Park and E. W. Burgess, Eds., *The City*. Chicago: University of Chicago Press, 1967, p. 55 (original work published in 1925).

imply change not in the character of the whole succeeding zone but only in particular sections of a zone. Unique topographical features such as rivers, lakes, and hilly areas influence the physical structure of any given city in various unforeseen ways, none of which neatly fit the uniform pattern of a set of concentric zones. In addition, transportation systems such as railroads, highways, and major streets and boulevards can interrupt the zonal pattern. These intersecting patterns of physical topography and zones of urban expansion tend to differentiate the city's land into a variety of small subzonal areas, which the Chicago sociologists called *natural areas* because they were assumed to be unplanned, natural products of urban growth and competition for land.

In the setting of the industrial city, major objects of competition include not only economic and social position, measured in terms of income and social status. For housing, for example, they also involve physical position in the community, including type and condition of surrounding housing and related facilities and access to central business, employment opportunities, and other important points in the community, such as health and educational facilities. The specific location of any particular natural area with regard to these factors significantly influences housing and land values. As a result, as population groups move into the city and among its different zones, they are sorted out in relation to their economic characteristics, in terms of what they can afford to pay in rental or other housing costs. At the same time, the cultural characteristics of the migrating individuals and groups also play an important role in the sorting process, with each natural area selecting from what Zorbaugh (1961) described as the "mobile competing stream of the city's population . . . the particular individuals pre-destined to it" by their cultural traits (p. 47). Thus, like competition, dominance, and succession, *spatial segregation* based on social and economic differences is perceived as a natural process. As Zorbaugh observed:

> The natural areas of the city tend to become distinct cultural areas as well—a "black belt" or a Harlem, a Little Italy, a Chinatown . . . a "stem" of the "hobo," a rooming-house world . . . or a "Greenwich Village," a "Gold Coast," and the like—each with its characteristic complex of institutions, customs, beliefs, standards of life, traditions, attitudes, sentiments, and interests. . . . Natural areas and natural cultural groups tend to coincide. (p. 47)

In short, in classic urban ecology the structure of the city is viewed as the product of the operation of the market in land interacting with the demographic characteristics of its various social and economic groups. The life and growth of the city are ultimately governed by supposedly unconscious *biotic processes of competition,* out of which there evolves in an unplanned, spontaneous way a spatial mosaic of culturally and economically segregated natural areas representing the placement of each group into its appropriate ecological place. As Burgess (Park & Burgess, 1925) notes, "This differentiation into natural economic and cultural groupings gives form and character to the city. For segregation offers the groups, and thereby the individuals who compose the groups, a place and a role in the total organization of city life" (p. 56). According to Park (1952), this is functional for the city as a whole, as well as for its constituent members, because it is assumed that in its particular niche each urban actor "will meet the least competition and contribute most to the life of the community" (p. 161).

Some Criticisms of Classic Ecology

Although urbanists have become skeptical of these highly optimistic claims today, this theory captured the imagination of urban students for over a generation, and some of its aspects still influence urban social science. Its considerable historical influence may be accounted largely to the fact that its method was empirical, focusing on observables, particularly the distribution of populations and land uses in urban space, and its theory was in good part social-Darwinian, echoing the themes of competitive social selection and organic natural evolution that were central to the intellectual milieu of laissez-faire capitalism in early 20th-century America. The theory finally went astray in its sharp separation of the biotic level of the community, viewed as a spatial ecology involving competitive economic relationships, from the cultural level, viewed as a system of shared communications and social order. The classic ecologists paid particular attention to the biotic level, for they observed that social life in the developing industrial cities of America was shaped significantly by competition for available resources, including land—a competition so intense that it could be compared with the struggle for existence in the animal world. Thus they theorized that the basic forces that decided the physical form and social structure of the community—its division of land as well as its division of labor—were the biotic processes of ecological and economic competition.

What is particularly interesting is that implicit in this theory, based as it is in the primacy of competitive processes, was the notion that competition was ultimately self-limiting. Through occupational specialization individuals and groups could establish a reasonably secure functional position in the division of labor; through segregation into different natural areas each could find a protected ecological position. In this way biotic competition for limited environmental resources led to long-term relationships of functional interdependence, or *symbiosis*, between individuals and among groups—a stable biotic equilibrium that can be understood as the ecological order of the community. Although biotic competition could be destructive to particular individuals or groups, the assumption was that such competition was functional for the community as a whole, imposing adaptive principles that were essential to the community's basic ecological order.

Yet this is not all. Emerging from this biotic-competitive level of the community, there ultimately arises a higher level of social order. At that level the classic ecologists identify a different set of organizing principles, in terms of mutual communication, the sharing of social norms, and the development of a consensual moral and political order. This is the cultural level of the community, which Park refers to as *society*. It is within this arena that social consensus and positive social control can be achieved. As Park (1952) observed:

Society is everywhere a control organization. Its function is to organize, integrate, and direct the energies resident in the individuals of which it is composed.
One might, perhaps, say that the function of society was everywhere to restrict
competition and by so doing bring about a more effective co-operation of the organic units of which society is composed. (p. 157)

Once the level of society emerges, it imposes itself as an "instrument of
direction and control upon the biotic substructure" (Park, 1952, p. 158). In
these observations Park was describing somewhat abstractly the emergence of
social institutions in early 20th-century America that reflected the rise of the
notion of a public interest that transcends the diverse private interests included in the community. These institutions—particularly those of government—
were gradually acquiring the power to regulate and restrict the competition
between private interests for the benefit of the larger community.

In spite of the obvious importance of the cultural level to understanding
the development of the modern urban community, for the classic ecologists
the proper focus of urban study remained fixed at the biotic or subcultural
level, below the level of human consciousness. In their view, the cultural level
with its speculations and assumptions about such intangibles as shared group
values and mentality was more appropriately the subject of social psychology.
Consequently, in attempting to separate sociology from social philosophy and
make it more "scientific," the classic ecologists chose to set aside basic issues
of social motivation that are involved in concepts such as social values,
interests, and purpose, which are vital to any thorough sociological analysis.
Their intention was to deal only with social facts such as the movement of
population, its geographic distribution, the physical location of its institutions
and facilities, and the like.

As a number of critics of this viewpoint made clear, if the cultural level is
in fact the level of consciously evolved social control, its operations and their
modifying effect on the biotic level of ecological competition cannot be
ignored without a serious loss of understanding of social process in the human
community. Furthermore, the division between biotic and cultural levels
could not be made realistically because "impersonal competitive relations as
defined by classical ecologists are so intertwined with personal cooperative
ones it is only by abstraction that we are able to separate one from the other"
(Hollingshead, 1961, p. 111). Indeed, as Berry and Kasarda (1977) noted in
their review of ecological theory, a series of critiques during the late 1930s and
1940s exposed a variety of shortcomings in the basic framework of classic
ecology. Besides "its muddled distinction between biotic and cultural ele-

ments," to which I have referred, they noted its "excessive reliance on competition as the basis of human organization, its total exclusion of cultural and motivational factors in explaining land use patterns, and the failure of its general structural concepts, such as concentric zonation and natural area, to hold up under comparative examination" (p. 6).

Confronted by this thoroughgoing critique, traditional human ecology tended to withdraw to a relatively limited set of objectives. After the 1930s urban ecologists increasingly narrowed their activities to performing empirically precise ecological and demographic studies, including careful studies of spatial distribution in urban communities of a variety of behavioral problems such as crime, delinquency, and mental disorders and the development of indices with which to statistically identify ecological variations among different social areas of the city (Rees, 1972; Shevky & Bell, 1955). In addition, demographic studies were made of several different types of population movement: of migration from rural to urban areas, of population shifts within cities, and of migration from cities to suburban areas with the end of World War II. Overall, there was a retreat from the ambitious early theoretical perspective of classic ecology, so that among urban ecologists "a majority shifted their attention to the narrow and somewhat unimaginative graphing and mapping of social data to uncover spatial (ecological) correlations" (Berry & Kasarda, 1977, p. 7).

CONTEMPORARY ECOLOGICAL THEORY

Under these circumstances, traditional ecological theory has in the course of the postwar period been replaced by the conceptual revision sometimes referred to as *neoclassical ecological theory* (Wilhelm, 1973). Hereafter it will be referred to as *contemporary ecological theory* or simply *contemporary ecology*. As I have noted, in the traditional ecological perspective, the city was viewed as the organic product of a general competition for resources among its various inhabitants. Contemporary ecology has modified the earlier one-sided focus on competition and has abandoned the distinction between a moral-political superstructure of the community that incorporates socially shared values and a biotic substructure shaped by processes of naked economic competition. The forces of competition are still at work, but the spotlight now shifts toward several other factors, primarily to the forces of technological innovation (such as automotive technology and electronic communication) that enable the integration of vastly enlarged urbanized areas and secondarily to cultural tastes and biases as expressed in the market for land both inside and outside the central city. The result is basically a technological-economic theory of urban development, modified by the influence of certain cultural and ideological factors, such as concern for maintaining ethnic homogeneity in local community life.[2]

Typical of the emphasis on technology is the contemporary ecologists' account of the evolution of metropolitan form in relation to late-19th- and early 20th-century developments in transportation, moving from the horse-drawn streetcar in the 1870s to the development of electric trolley lines and electric railways in the 1880s and 1890s. By the early 1900s this had produced a pattern of urban development essentially limited to the corridors adjacent to these rapid transit lines. It was only the introduction of the automobile that overcame the limited radial pattern of concentrated development along the transit lines, incorporating the whole suburban zone into the orbit of the growing metropolis.

By the 1920s, however, the automotive revolution made possible the massive dispersal to all parts of the suburbs of central city populations along with industrial and commercial facilities. They now were free to respond to *push factors* of "obsolete inner-city structures, lack of inexpensive space for expansion, increasing taxes" (Kasarda, 1978, pp. 36-37), and the like and to *pull factors* connected with access to relatively cheap, abundant residential and commercial space, large modern industrial production facilities, and reduced transportation costs to manufacturers and retailers locating near suburban expressways. Overall, "changing modes of transportation and production technology altered central cities' locational advantages" (p. 36) and produced a powerful logic of expansion into the suburbs and even beyond into other regions.

This growth-oriented spatial perspective can be related to a basic concept in contemporary ecology, the notion of the *ecological system*. In part, this concept attempts to respond to the enlargement in the scale of urbanization in the metropolitan era by replacing the traditional ecological concept of the physically contained urban community modeled on the industrial city. During the period of rapid suburbanization and metropolitan expansion following World War II, Duncan (1964) postulated the notion of an *ecological complex* that is an evolving social spatial system consisting of the three major classic ecological variables: population, environment, and technology, to which he added a fourth, social organization. Because these four variables all interact with and condition one another, they are referred to as an ecological system or ecosystem.

Growth or expansion of the ecosystem is a cumulative process, which today most frequently begins with changes in technology. In particular, an accumulation of advances in scientific knowledge and technology makes possible new uses of the environment and its resources. This increases the environment's carrying capacity and allows for an expansion of the population that applies the new technologies. That in turn should lead to adjustments in social organization, such as new administrative arrangements and changes in the division of labor in an expanded territory of settlement. On completion of these cumulative steps, the way is opened to the next cycle of expansion (Duncan, 1964).

Although theoretically all four variables in the ecological system mutually and reciprocally condition one another, social organization is clearly the dependent variable; the others are the active ecological variables to which the community adjusts its organization. In a chapter on "Human Ecology and Population Studies," Duncan (1959) noted that among the "significant assumptions" that the ecologist makes about social organization is that it "must be adapted to the conditions confronting a population—including the character of the environment, the size and composition of the population itself, and the repertory of techniques at its command" (pp. 682-683). Since, as the contemporary ecologists themselves note, the environment "in the modern industrialized city . . . has essentially been created by man's application of his advanced technology" (Berry & Kasarda, 1977, p. 15), we have the paradox of a human population being required to adapt itself not only to its own environmental creations but also to its own technology, "the repertory of techniques" that are presumably "at its command" (Duncan, 1959, pp. 682-683).

This is a serious practical and theoretical problem for all those concerned with the development of a modern urban society. It suggests that at the very heart of the urban development process there is an absence of active control of the direction of that process by the community's population. It is the forces of technology and the environment created by them to which the community's social organization and hence its very way of life must be adapted, rather than the reverse! Yet contemporary ecological theory offers little insight into this paradox because it does not examine closely the urban-related operations of those social structures that bear most relevantly on control of the urban development process. In short, contemporary ecology does not focus on the vital role of the institutions of today's government and economy in shaping the social-physical environment of urban and metropolitan populations. The emerging organizational structure and policy implications of these institutions will be discussed at length in later chapters.

METROPOLITAN EXPANSION AND URBAN DEVELOPMENT

In contemporary ecology, *ecological expansion* is defined as "a process of cumulative change, whereby growth of a social system is matched by a development of organizational functions to insure integration and coordination of activities and relationships throughout the expanded system" (Berry & Kasarda, 1977, p. 15). Ecological expansion presumes, therefore, not merely quantitative growth in the system's territory and population. It also requires a growth in organizational and integrative capacity, which neoecologists associate with the centralizing of administrative control functions. There is considerable evidence that an increase in centralization of administrative functions indeed has occurred in the postwar period in both the economy and in government. The question is, however, to what extent has that development led to an integrated metropolitan community?

If an ideal process of metropolitan expansion could be arranged, it would involve the balancing of what is referred to by contemporary ecologists as the centrifugal and centripetal movements involved in expansion. *Centrifugal movements* are those that have taken population, commerce, and industry out of the central cities. Such out-migrations often have contributed to the deindustrialization of formerly industrial cities. Combined with the postwar influx of low-skilled populations that have high dependence on public services, these centrifugal movements have weakened the budgetary bases of many of the larger older industrial era cities and led them into fiscal crisis.

Centripetal movements, according to the neoecologists, tend to produce a concentration of administrative activities in central cities as well as an array of professional, technical, financial, and related business services (Berry & Kasarda, 1977). The result in some large cities that have been able to build up these new activities has been not simply a loss of industrial base but also the first stages of development of a postindustrial city offering an expanded range of specialized business and professional services. Overall then, this has been a situation of uneven development, bringing benefits and opportunities to some segments of these cities, along with displacement and hardship to others. The mere fact of the growth and elaboration of the central city white-collar workforce has not assured a successful transition in which all the elements of the metropolitan area are integrated into a unified community. Although the centripetal movements of metropolitan development have brought a concentration of administrative and professional specialists to the central city, this does not mean that their capacities have been used to integrate fully the surrounding metropolis even in its economic aspects, much less socially and politically.

The first generation of neoclassical ecologists described the process of ecological expansion as if its centrifugal and centripetal aspects were intrinsically self-balancing. Writing in the early phase of postwar metropolitan expansion, Hawley (1950) defined *expansion* as a phenomenon that "involves centrifugal and centripetal movements. Centrifugal movements are the process by which new lands and populations are incorporated into a single organization. The centripetal movements make possible a sufficient development of the center to maintain integration and coordination over the expanding complex of relationships" (p. 369). In particular, the factor of social organization takes on added importance in this context. Hawley's definition assumed that the spatially expanded environment would be bound together by integrative organizational developments in the central city.

Later analysis of data from the 1960 census appeared to support this assumption. Berry and Kasarda (1977) constructed a measure of *organizational development* representing development of administrative capacity in the central city, viewed as the organizational nucleus of the metropolitan area. The measure they adopted was analogous to that of *administrative intensity* within a formal organization, which divides the number of personnel performing administrative or related functions by the total number of personnel. This was applied to seven categories of employees working in the central city in administrative or support functions related to business or public administration. The categories include managerial functions; professional and technical functions; clerical and other support functions; transportation and public utilities; finance, insurance, and real estate; business and repair services; and public administration. The number of employees in each category was divided by the central city population size, giving seven detailed measures of organizational development in the city. Similar measurements were made for organizational development in the same categories in the surrounding suburban ring.

Berry and Kasarda (1977) found that all the indices of organizational development in central cities tended to increase with growth in the population of the suburban rings. Overall, they concluded that central cities are more developed significantly in their administrative functions than are their suburban rings and that increases in suburban ring population "have been matched with a development of organizational functions" in their central cities (p. 209).

This study of 1960 census data appeared to support the assumption that the central city would increase its administrative capacities to integrate the expanding metropolitan area. However, the expansion of administrative capacities can have various applications and might well serve purposes other than metropolitan integration, such as integration of national or global administrative networks, as in the case of cities such as Washington or New York. In addition, it should be noted that the data studied reflect metropolitan area relations less than two decades after World War II. Hence they refer to the experience of an early phase of postwar metropolitan expansion when the central city still exercised relative dominance over its metropolitan environs.

After another two decades, by the late 1970s the "strains of metropolitan expansion" already were beginning to show unmistakably. One problem is that the spatial reorganization of the metropolitan area, which was so much the focus of concern of the early neoclassical ecologists, has not been matched by its political reorganization. Although the outlying villages, towns, and small cities have been incorporated economically and to some degree socially into the web of the metropolitan community, they remain politically autonomous units free of any legal obligations to a central government of the area. Consequently, the typical metropolitan area is today, with regard to political organization, a jumbled mix of different overlapping governments. Overall, the net effect has been "fragmented taxing powers, public service inefficiencies, conflicting public policies, and administrative impotence in dealing with many of today's problems that have become metropolitan in scope" (Kasarda, 1978, p. 40).

As a consequence, contemporary ecologists no longer tend to assume any intrinsic balance between the centrifugal and centripetal movements that have shaped metropolitan expansion. Instead, they tend to treat the relationship as problematic. As Kasarda (1978) observed, "We are arguing that the centrifugal and centripetal movements inherent in metropolitan expansion are at the root of the current economic and fiscal problems plaguing our large central cities" (p. 44). Instead of complementary development, centrifugal and centripetal movements have resulted in several basic mismatches. First was the *fiscal mismatch* produced by the suburban out-migration of middle- and upper-income taxpayers simultaneous with the influx of large numbers of low-income families and individuals. Second, the *employment mismatch* was produced by suburbanizing large numbers of blue-collar jobs, thereby placing them out of reach of inner-city dwellers, while bringing numerous white-collar suburbanites into the city for office work. Finally, the *skills mismatch* has resulted in the placement of corporate office complexes in central city downtown areas amid residential populations whose skills fail to match the new jobs.

Today after more than a generation of continuing decentralization, the dispersion of once basic elements of the industrial era central city has led to the decline of many large cities from the dominant core role that they played from the 1920s to the 1970s. Now "core-dominated concentration is on the wane; the multinode, multiconnection system is the rule" (Berry & Kasarda, 1977, p. 267). The emergence of suburban communities that are increasingly self-sufficient, incorporating significant commercial and industrial facilities together with large diversified employment bases, means the central city is often now merely one urban node among many in the metropolitan area (Muller, 1981).

URBAN DEVELOPMENT IN ECOLOGICAL PERSPECTIVE

The foregoing provides only part of the picture of urban expansion today. The situation has gone beyond the expansion of the central city into the suburban zone and beyond the creation of multiple points of urban development within that zone. Contemporary ecosystem expansion has moved into an era of major interregional and even international shifts of facilities, jobs, and industrial development. What is entailed is no longer merely the breakdown of the distinction between urban and rural in the environs of the older industrialized cities, particularly in the northeastern urban-industrial heartland. That was a consequence of the large-scale suburbanization that characterized America in the decades following World War II, establishing the setting for the postwar emergence of neoclassical ecology. Today the issues have moved on to include the breakdown of the historic distinction between the traditional urban-industrial heartland as the dominant ecological region and the formerly rural hinterland that is now rapidly urbanizing, reflecting the effects of ecosystem expansion on urban development on a national scale. As a result, contemporary ecological theory has become a theory of the reorganization of the space-economy of the urban complex at the level of a national society driven by postindustrial forces of change and of the impact of this reorganization not only on older industrial central cities but on the entire industrial heartland in relation to outlying regions of the nation.

In this connection, the most significant postwar trend in the redistribution of urban development between regions has been the net shift of population and industrial growth from the former urban industrial heartland of the North, first to the West and then, since the 1970s, increasingly to the South. In part, the southern regional gains can be explained in terms of an accumulation of changes in recreational lifestyles, more cosmopolitan cultural attitudes, lower costs of living and of land, and improvements in personal income and in consumer services in the region. But by far the most important factor, on which many of these other changes were themselves contingent, has been the emergence of the South as "the major industrial growth pole of the U.S. economy" (Kasarda, 1980, p. 376).

What is it that has made the South so attractive as an industrial growth pole? Consistent with the contemporary ecologists' emphasis on transportation and production-related factors, the primary pull factors include new interstate highways that have made the South widely accessible to national markets, plus changing relative costs among energy sources that have made its oil and natural gas economically attractive in a period of conversion from coal. In addition, lower wage rates in the strongly antiunion South have acted as a further inducement to regional industrial investment and growth, together with strong probusiness attitudes among southern communities and their public officials. The result has been the development of a conscious corporate policy of locating new manufacturing growth in this region. Major push factors cited as responsible for postwar industrial influx include "congestion, powerful unions, high land costs, high taxes," and other "negative externalities" of the crowded metropolitan areas of the northeastern and north central regions of the nation (Kasarda, 1980, p. 377).

What are some of the policy implications of current redistributional trends? Kasarda (1980) reflected on "the push and pull factors involved in the remarkable job growth in the South" (p. 389). In essence what he recommended to policy makers is an application of the southern model to the old industrial heartland in the North. This will require development of a favorable business climate through the modification of local regulatory policies that tend to discourage business creation or expansion. Also suggested is an educational campaign to help develop probusiness attitudes among local officials and residents. Other points include the elimination of inessential services and programs to ease municipal fiscal strains and to permit increased spending on programs that can enhance the competitive situation of older industrial cities, such as improvement of public schools and maintenance of essential urban public infrastructure, from streets to sewage systems. Overall, the basic thrust is aimed at establishing that a national urban policy should "work *with* the forces of redistribution rather than fruitlessly attempting to reverse them" because "large, dense concentrations of people and firms have become technologically obsolete" (p. 393).

In this perspective, if there is hope for the older central cities, it lies in adapting to postwar technological changes, not in denying those changes by trying to lure back to the cities the manufacturing industries that have left them in that period. The best prospects for the large central cities lie in their continuing transformation to postindustrial centers of service consumption rather than in attempts to regain their status as centers of industrial production. Accordingly, the neoecologists recommend that such cities should continue to encourage the growth of administrative and professional jobs in the central business areas as a way of providing advanced services to expanding business complexes throughout the metropolis. In addition, the cities should focus on revitalizing their core areas into "culturally rich, architecturally exciting magnets for conventions, tourism and leisure-time pursuits" (Kasarda, 1980, p. 393). Finally the older cities should support the restoration of historic neighborhoods adjacent to the central business districts in order to increase the appeal of these core areas and provide conveniently located housing for CBD employees.

The logic of this position is thus to use public policy and public resources to support the transformation of the central city from an industrial site occupied by lower class blue-collar populations to a postindustrial office and cultural-leisure center whose residents, particularly in areas near the CBD, preferably are to be white-collar administrative, professional, and clerical workers. However, a serious problem exists in the implementation of this approach. What is to become of the large residual working-class and minority populations who reside in broad areas of the inner cities, including the close-in areas adjacent to the CBDs?

Kasarda's (1980) response is two pronged. First, national urban policy should act to eliminate discriminatory zoning and real estate barriers that deny housing opportunities to minority and lower income persons near blue-collar job complexes in suburban and nonmetropolitan areas. It also should help develop additional low-income housing near expanding job sites outside central cities. Second, it should "support intensive, up-to-date technical training programs that will provide all those desiring employment, with appropriate skills" (p. 395).

Although these suggestions are eminently reasonable when viewed within the assumptions of the ecological perspective, they may not take fully into account a number of significant social and political realities of the current situation. For one, there exist powerful barriers to the in-migration of Black or low-income people in most suburban areas in the nation. Among the numerous obstacles that exist to open entry into residential suburbs, some are legal or de jure and serve to legitimate segregation on an income basis. Such mechanisms are prevalent, well known, and extremely difficult to attack effectively. Together they comprise a system of exclusionary zoning, which makes it possible for the community, using its powers of land and building control, to "guarantee that the sale and rental prices of properties are not affordable by low- and moderate-income individuals of any social group" (Muller, 1981, p. 95).

At best, then, most suburban communities are effectively closed to all but a relatively narrow spectrum of upper middle-income minority households. However, even if all de jure exclusionary mechanisms could be removed, numerous unofficial or de facto income and racial barriers would remain in place. These include the "steering" by real estate people of buyers to different residential areas by race, clandestine observance of protective covenants and "gentlemen's agreements" to exclude members of designated racial or ethnic groups from particular areas, and the denial by lending institutions of mortgage or improvement loans to areas that house substantial minority or low-income populations, a practice known as *redlining* (Muller, 1981).

Most likely the de facto obstacles to suburban integration will continue in the absence of a thorough, sustained nationwide legal and educational effort beginning at the federal level, the likes of which has never been attempted by any administration, Republican or Democratic. But even if the suburbs could be opened up fully, there remains another significant force against integration of minority groups, and specifically the Black central city population, into the suburbs. That is the considerable resistance that exists within the Black community itself to suburban integration.

In part, this resistance is a product of long experience in encountering suburban racial barriers, which range from the mechanisms already discussed to open verbal abuse and physical violence for many of those who have acted in the past as pioneers in White suburban areas. The result is disillusionment and cynicism in facing what amounts to a dual housing market with separate points of residential entry, pricing, and lending practices, differentiated by race.

But another aspect of the internal resistance to suburban integration is more political in nature. That is, there is an aversion among Black political and institutional leaders in general to what they perceive as the potential scatteration of central city Black communities to the detriment of their respective institutional interests in churches, businesses, and various neighborhood organizations that would be losing clientele and to the possible diminution of their political influence in the central city. Thus, in response to a relatively small Areawide Housing Opportunities Program (AHOP) sponsored by the Department of Housing and Urban Development (HUD) in the Baltimore metropolitan area in the early 1980s, a group of local activists, who referred to themselves as the Baltimore Coalition to Save Urban Communities, spoke with area officials and held a press conference to protest AHOP. The report by Constable (1980) in the *Baltimore Sun* noted that:

P. M. Smith, a Baltimore lawyer, and housing activists in a dozen other major cities say the program is part of a subtle government plot to diffuse the power of poor and black urbanites by scattering them in the countryside while opening up newly fashionable inner cities to the white middle class. (p. B1)[3]

Whether the allegations concerning this particular program are valid or not, the fact remains that such perceptions do exist and act as a restraining element against the suburbanization of Black households. They are grounded in many decades of rejection, disillusionment, and distrust and will not be quickly or easily overcome.

Thus the specter of intergroup perceptions, politics, and power comes to haunt the rather mechanistic ecological theory, in which urban change proceeds on the basis of a seemingly inexorable, spatially expansive model. In that model, market choice and technological innovation join to drive the process of urban development ever outward from its original bases in the older industrial central cities of America. To stabilize the process and make it optimal for adapting the older cities to their new role as service and administrative centers requires the freeing up of the areas next to the CBDs for further office expansion or for development as renewed residential, commercial, or recreational facilities. This often has meant the displacement of adjacent low-income residential populations. As Kasarda (1980) himself observed, "Large concentrations of the poor around the CBDs pose serious negative externalities for the cities by discouraging shoppers and tourists from using core facilities at night and by substantially increasing the risk of new capital investment on CBD peripheries" (p. 394).

In the ecological perspective, the logic of central city redevelopment in the wake of metropolitan expansion is clear. Indeed, if substantial suburbanization of lower income minorities could be achieved with decent housing and upgraded employment for the new suburbanites, there might well be benefits for all involved. But the realities of majority biases and minority fears intrude and bring to mind the missing dimension in the ecological perspective—the dimension of interests, values, politics, and power—in the reshaping of contemporary cities whose roots lie in the industrial past.

NOTES

1. Of the 32.8 million European immigrants to the United States from 1830 to 1920, 23.4 million or 71% arrived after 1880 (Ward, 1971, p. 53). In addition to their distinctive language and religious differences, the members of the new migration were much more highly visible to

established residents particularly because they tended to move overwhelmingly into cities, rather than distributing evenly between city and countryside as had, for example, the Scandinavians of the earlier migration.

2. See Berry and Kasarda (1977), who present an explanation of the continuing existence of residential segregation in contemporary America in terms of the rise of a "mosaic culture" based on intensification of subcultural identities and residential self-segregation of different subcultural groups (p. 80). The development of this mosaic culture based on subcultural divisions of race, class, and lifestyle is related by the authors to postwar ecological trends of decentralization from central cities and the creation of socially fractionated metropolitan areas. These trends are interpreted as manifestations of a new social dynamic (associated with an ideology of cultural pluralism), one of whose major aspects has been the phenomenon of White flight to the suburbs "as minorities move toward majority status in the city center" (p. 266).

3. This article passage is used courtesy of The Baltimore Sun Company, © 1993, *The Baltimore Sun.*

[2]

The Marxist Perspective

In Chapter 1, we reviewed the evolving perspective of ecological theory. This theory, rooted in the notion of a human community that is an organically integrated ecological system, has established itself as a central theory in American urban studies and thus as a major basis for contemporary urban planning. However, in the wake of the urban turmoil of the 1960s, there was growing recognition that many American cities in fact had failed to socially integrate significant segments of their populations and that this ultimately contributed to an atmosphere of sharp political and economic conflict. Since that time, several conceptual alternatives have emerged to challenge the relatively conflict-free mainstream ecological theory. The major challenges derive from the marxist theory of urban development, focusing on the conflict of economic interests that are conceived to lie at the root of such development.[1]

For the marxists the critical factors that underlie contemporary urban development and its problems are connected with a fundamental economic process that they refer to as the process of *capital accumulation.* In their view it is the capital accumulation process, rather than technological or related developments emphasized by the ecologists, that explains the trends toward metropolitan expansion that have affected so seriously the situation of older central cities. Nor are market forces by themselves—responding to consumer preferences for cheaper land, lower taxes, or greener spaces—the basis for the great exurban migrations of this century. Above all, what has changed the shape of urban America has been the economic activities of powerful entities such as corporate business firms interlocked with financial institutions and acting with the support of government agencies, shifting investments in manufacturing and commerce to the suburbs and concentrating investments in administrative functions and services in the central cities.[2]

URBAN CHANGE AND CAPITAL ACCUMULATION

Tracing the key elements in this process will indicate how different the marxist perspective is from that of the contemporary ecologists, yet how the two approaches nominally overlap at several important points. In the marxist vocabulary, capital accumulation is the process of capitalist profit making and implies that the profits involved will be used to accumulate yet more capital for further profit-making ventures. Capital accumulation is an expansive economic process capable of introducing important changes in society, not only in the size of its wealth but also in its economic class relations and in the character of its cities and surrounding regions.

The perspective of marxism is congenial to the study of processes of socioeconomic change. One of its central tenets is that no social structure is fixed for all time. Specifically, as the mode of production (the economic system) changes, so too will all important aspects of a society. In particular, both the process of capital accumulation and the form of urbanism will vary with the prevailing mode of production. Moreover, there is evidence that capital accumulation and urbanism are closely connected, for historically it can be shown that the city plays a central role in the accumulation process.

The first of the three major aspects of the capitalist accumulation process (Mandel, 1970) is the economic *production* of goods or services such that the value of the product is greater than the price of labor plus the cost of materials used in production. Because in the marxist perspective all new values are created only by human labor, the net increment in value introduced in the production process (surplus value) is traceable to the efforts of labor and amounts to the difference between labor's full value and its current wage, reflecting the prevailing degree of exploitation of labor during any given period. Instead of the surplus value being returned to its producers, it is taken (appropriated) by the owners of the firm upon its production and economically realized by them when the product is sold. Thus *appropriation* of surplus value by the capitalists presupposes the existence of a market for the product of the firm. It should be noted that cities, as concentrations of settled population and as geographic central places, historically have served to provide access to a wide variety of markets within and beyond their own boundaries.

Under capitalism cities come to play a key role in the process of appropriation of society's economic surplus.

Now, appropriation is only part of the accumulation process. Having appropriated the profit from the sale of the product, the capitalists now typically look for ways to increase it. Thus, if the future market looks favorable, they reinvest in the firm. Should this *reinvestment* pay off by increasing the productivity of the firm, for example, by installation of new technology, the owners now obtain more output per hour from the workforce, thus expanding the profits annually. This increases available capital, reflected in the increased amount of money they now have with which to operate the plant and make new investments in it. In part the available capital is also made up of credit on which they can draw from the banks, based on the previous successful record of capital accumulation. Overall, this specifies three major stages of the accumulation process: production; marketing and profit taking (appropriation of surplus value); and reinvestment leading to expanded production, beginning the cycle of accumulation again. This outline of the accumulation process can be related now to some of the historical functions and forms of American cities.

The first stages of American urbanization occurred during the period when America was predominantly an agricultural country and variously took the form of frontier towns, market towns in settled rural areas, and commercial cities that developed mainly around the bustling seaports of the East Coast. These cities dominated the urban scene from colonial days almost to the Civil War. However, during this period "most of the surplus which cities helped urban elites to accumulate was not actually produced in the cities themselves; rather the bulk of it was produced in, and extracted from, the countryside" (O'Donnell, 1977, p. 93). Consequently, the commercial cities specialized in the second and third stages of the accumulation process, serving as *centers of commercial exchange,* primarily for agricultural commodities produced in the rural hinterlands, and as *centers of finance,* providing banking and credit services necessary to conduct commercial ventures.

Although such cities often contained sizable numbers of craftspeople who produced handworked items for the local market, production was typically a family affair, conducted on a small scale (*petty commodity production*). Handicraft production generally was decentralized, taking place in the workshops and homes of independent craftspeople and artisans, with home and workplace often in the same physical structure. For the most part, capital accumulation in the commercial cities was not centered on handicrafts, which were too scattered and limited in volume for large-scale commerce. Most profits were made instead in the process of buying and selling commodities such as cotton or wheat in bulk quantities brought in by ship or riverboat, that is, in the exchange of items produced elsewhere to be sold in the city and then distributed from there often to be consumed at some other location. The commercial city was essentially a point of exchange for the products of agriculture and only secondarily for handicrafts, and capital was invested and accumulated in connection with the exchange of goods rather than with their production.

By the 1830s merchant capitalists, seeing the possibility of a new source of gain in harvesting the profits of a larger volume of handicraft production, introduced the *putting out system,* by which raw materials were distributed to households of semiskilled workers and the finished goods, such as textiles, were then collected and sold by the merchant entrepreneurs. However, this system still allowed the household producers to regulate their own production, in which circumstances "they tended to limit their output and set aside only a small amount of capital for investment" (O'Donnell, 1977, p. 94). The results were disappointing to ambitious merchant entrepreneurs.

The putting out system was, after all, basically the system of petty commodity production with the addition of centralized commercial marketing. As such, it could not transcend its limitations and managed to expand only very slowly. In its place, after about a generation of trying to work with the putting out system, venturesome American entrepreneurs set about organizing a new system of production, which is known today as *industrialism.* The essence of this system consisted of concentrating workers at a central point of production and working them, closely supervised, longer hours than they normally would at home. This became the basis of the factory system in America, involving long hours, extremely strict work discipline, and unquestioned control of the organization and pace of work by the owner-entrepreneur. Through control of the work process, "capitalists were able to increase greatly the total volume of production, the amount of surplus value produced, and the quantity of capital that was accumulated" (O'Donnell, 1977, pp. 94-95).

By the end of the century American society had been launched decisively on the road to an economic system in which capital accumulation took place predominantly within a hierarchically organized context of factory production. Now the main source of capital accumulation became the making of profits by the process of industrial production, "the direct manufacture of the commodities that they exchanged on the market," rather than the collection of commodities from other sources for exchange, as in the commercial city

(Gordon, 1984, p. 30). According to the marxists, it was to be expected that these changes in the mode of production would be reflected in the structure of the city, most visibly in the use of land, as American cities changed from commercial to industrial centers.

Land use in the commercial city, as in other preindustrial cities, was loosely patterned although not unstructured (Jones, 1966). In general, the docks determined the primary focus of economic activity. Bordering on them could be found a broad zone of mixed land uses related to commerce and exchange: "Warehouses and countinghouses, the establishments of greatmerchants, and the retail outlets of petty tradesmen, the taverns and grogshops all crowded close to the waterfront, and the longshoremen, hustlers, clerks, shiphandlers, sail-makers, and coopers lived nearby" (Walker, 1978, p. 176). Most of the rest of the city was given over to residences, with the exception of a small area of administrative uses, such as the town hall and other public buildings, centrally located between the commercial district and the town houses of the leading citizens. Nearby also were the major churches of the city.

The division of residential land was essentially opposite to that which later characterized the industrial city. It was the poor who lived at the edges of the commercial city; the affluent resided at the center, with the artisans and petty tradespeople located between the two (Ward, 1971). Wealthy merchants, professionals, and public officials occupied the most central locations of the inner residential area, within easy walking distance of both the commercial zone and the administrative facilities of the city. Yet overall there was a wide mixture of people from various backgrounds "interspersed throughout the central city districts, with little obvious socioeconomic residential segregation" (Gordon, 1984, p. 27). In fact, in the fluid circumstances of the commercial city, there was neither the clear-cut social class structure, nor the distinctive spatial segregation of land uses, which developed later in the natural areas of the industrial city. Only the poor, the Blacks, and the itinerant laborers were set off clearly from the rest of the community, living in shanties at the edge of the city. If there was a general rule regarding location in the commercial city, it was that the farther from the center, the less desirable was the location.

Nevertheless, there was as yet no sharply delineated separation of land uses. The phenomenon of ecological dominance of major land uses over their surrounding areas tended to be limited to the wharf and warehouse sections of the commercial zone and was not typical in the more diversified residential zone of the city. With the coming of the industrial city, these patterns of urban land use changed dramatically.

In this next phase of development of the urban community in America, the physical structure of the early industrial city paid testimony to the central role of the factory system. As Mumford (1961) observed, "The main elements in the new urban complex were the factory, the railroad and the slum. . . . The factory became the nucleus of the new organism. Every other detail of life was subordinate to it" (p. 458). The factory district often grew up at the periphery of the earlier developed central business and administrative district and along the railway lines leading into the city's central areas. Adjacent to this district there emerged distinctive areas of working-class housing. In order to locate the large and rapidly growing industrial workforce within walking distance of the factory area, workers' housing was crowded together in row houses or tenements, typically segregated from the better residential neighborhoods of the middle and upper classes. In this way, the foundations were laid for the sprawling inner-city slums of the mature industrial city. At the same time, the more affluent classes began to withdraw from the central areas of the city at an increasing pace, locating themselves "in concentric rings moving along the transport spokes radiating from the center" (Gordon, 1984, p. 36). With this the outlines of the familiar concentric zonal pattern of the industrial city were established, effectively reversing the spatial relations of the commercial city.

In the industrial city, the ecological dominance of the central business and industrial districts was clearly reflected in the patterning of the surrounding residential areas, which adapted themselves to the functional and spatial requirements of the central zone. This much is apparent whether viewed from the ecological or marxist perspective. However, for the marxists, the main point is that industrialization did not only change the spatial forms charac-teristic of the commercial city, it reshaped the city's fundamental economic and social relationships. What underlay these changes was the power of the capitalist class to centralize the means of production and through this to change the occupational structure and residential pattern of the urban workforce.

Thus the issue of dominance is a real one for the marxists; however, unlike the ecologists, their usage of the concept does not connote spatial dominance, as in the dominance of central land uses over peripheral ones. In the marxist perspective, it concerns *class dominance,* the social-economic dominance of one class over another in deciding the basic structure and content of the space economy of the city. For them, the spatial structure of the city is not simply an incidental result of technological change. Instead it is a consequence of purposive action by members of the capitalist class with an eye to stimulating the expansion of capital accumulation while maintaining their own dominant socioeconomic position. More generally, the power to shape the uses of space

within and outside the city is a function of the power to dominate class relationships both politically and economically. This political-economic power did not
end with the transformation of the commercial city to an industrial city; it persists
into the present and is a matter of central concern in the contemporary marxist
perspective, which views itself as a "political economy" of urban development.

An important point of departure for the work of marxist students of
urbanism is the question of the historical use of spatial strategies by the
capitalist class in its efforts to acquire and maintain control over the working
class at the workplace and in public life. During the industrial era, for example,
concern about maintaining control appears to have affected basic locational
choices regarding the urban placement of factory concentrations. Gordon
(1984) argued that by the decade after the Civil War, it would have been
technologically at least as efficient to locate major factory complexes in
medium-sized cities as in large cities because the railroad network by then
provided most medium-sized cities with access to adequate supplies of coal
and other raw materials needed for manufacturing. In addition, important
technological innovations were by that time sufficiently widespread that
industrialists everywhere had access to them.

Yet, following the Civil War, manufacturing became concentrated increasingly in a limited number of large cities. This can be accounted for not on the
basis of any special technological advantage enjoyed by larger cities but in
terms of the difference in social relations between the working and middle
classes in medium- and large-sized cities. Stratification was more clear-cut in
the larger cities where the working class was more likely to be segregated in
its own residential slums. Absence of sympathy for workers from the middle
and upper classes in big cities meant lack of support among the most influential segments of public opinion with regard to strikes and other labor
disputes and gave the employers a freer hand in dealing with labor unrest
(Gordon, 1984).

Physical segregation in big-city slums seems to have been applied then as a
strategy for political neutralization and social control of the first generations
of American industrial workers, dissatisfied with low wages and the rigid
hierarchic discipline of the new industrial system of production. However,
segregation of the laboring class proved to be only a temporary solution. By
the last decades of the 19th century, strikes and demonstrations had begun to
spread throughout industrial cities until labor agitation pervaded the urban
industrial environment.

Parallel with these developments in labor organization and militancy, American business entered a period of serious economic crisis in the early 1890s.

The structure of business was still highly competitive, with many firms engaging in price wars and other forms of cutthroat competition. These practices were basically self-defeating because they led to serious instability throughout the economic system. This instability found expression in the deep economic decline of the 1890s, in which many marginal firms went under. In these circumstances only the strongest could survive, and many smaller firms were swallowed up in a great wave of business mergers that continued from 1898 to 1903. These mergers introduced an era of major shifts from a largely competitive capitalist system to a more monopolistic type of capitalism. Over the next two decades a series of mergers and takeovers resulted in the concentration of capital in large corporate units whose great resources gave them unprecedented capacity to control both their workforces and their markets. In this way, industrial capitalism made the transition to corporate capitalism, whose leading units were giant corporations that sought economic stability and predictability in place of the uncertainties of free competition.

THE CORPORATE CITY AND
METROPOLITAN DEVELOPMENT

All of this was reflected in the shaping of a new city that marxist urbanists refer to as the *corporate city*. Until the very last years of the century, manufacturing continued to concentrate in the largest central cities. Then with the turn of the century, manufacturing began leaving the central city for the suburban rings. From 1899 to 1909 manufacturing employment in suburban rings rose by almost 100%, but rose only about 40% in central cities (Gordon, 1984, p. 40). In part, this reflected the fact that with added resources gained from mergers, major corporations developed sizable industrial tracts—and in some cases, what amounted to new outlying manufacturing towns or industrial satellites—in suburban areas surrounding major cities of the Northeast.

With this, an important step was taken toward significant expansion of the industrial city beyond its original bounds because many of its residents tended to follow the resulting new job opportunities with moves either to the city's outermost residential areas or out into the suburbs altogether (Kain, 1970).[3] In the ecological perspective, the decentralization of manufacturing that initiates the overall process is interpreted typically as a product of technological changes (such as the introduction of the motor truck to haul industrial freight) or a response to market signals (such as increases in central city land costs). In reply, marxists observe that the truck was not a viable substitute for railroad freight transport until the 1920s and that there is little significant evidence of rising urban land costs in the period prior to World War I (Gordon, 1984). Instead, they point to other concerns of industrial management having more to do with increasing difficulty in controlling the urban industrial labor force than with technical or market considerations.

In general, for the marxists matters of socioeconomic control and power are the major factors that decide patterns of urban development. They tend to be skeptical of locational explanations based primarily on technical considerations such as increased efficiency, even though they recognize their operation. From the marxist perspective, the dispersion of manufacturing facilities beyond central city boundaries was primarily a corporate response to labor unrest that had gripped the cities since the 1880s. As the president of a contracting firm in Chicago observed around the turn of the century:

> All these controversies and strikes that we have had here for some years have . . . prevented outsiders from coming in here and investing their capital. . . . It has drawn the manufacturers away from the city, because they are afraid their men will get into trouble and get into strikes. . . . The result is, all around Chicago for forty or fifty miles, the smaller towns are getting these manufacturing plants. (U.S. Industrial Commission, 1900-1902, vol. 8, p. 415)

The move was made by large corporations as soon as they had acquired sufficient capital assets to do so following the consolidating mergers of the late 1890s. Half a century earlier, the highly concentrated city was a product of the needs of capital to bring together a workforce in an environment where the organization and pace of the labor process could be controlled. Now at the turn of the 20th century, the first steps in reshaping the form of urban settlement toward a deconcentrated city by dispersing its industrial facilities were also a function of the desire of capital to control its workforce.

Suburbanization was not a new process; upper-income residents had been moving out of central cities since the late 19th century. In turn, cities played a game of catch-up to subsequently annex the new residential suburbs that thus were created. However, the rules of the game changed once the big corporations became involved significantly in it. The annexation of suburban residential areas was an established practice until the very end of the 19th century. Then the pace of annexation suddenly declined. This was not the result of an abrupt change in consumer preferences for suburban residential locations. The new element in the situation was that now manufacturing industry also had begun leaving the central city in order to find more favorable location with regard to a variety of concerns, including rising city taxes as well as difficult urban labor relations. The generous resources of industry provided it a degree of influence with state legislators that finally made possible the reversal of the policy of suburban annexation by central cities (Gordon, 1984). In the long run, the loss of the power to annex meant that continuing development of politically independent suburbs would lead to a proliferation of multiple governments in older metropolitan areas.

By the 1920s a broad-based constellation of economic interests had taken shape with regard to the development of suburban areas of major cities. In addition to industrial and commercial sectors of capital with direct stakes in decentralization, the auto and rubber manufacturers, oil companies, and highway construction industry, all of which shared an interest in fostering automotive transportation, became strong proponents of suburban road construction and supporters of growing dependence on the automobile that it encouraged.

Marxists observe that during the 1920s the interplay of these various forces produced an expanding process of suburbanization that was an important aspect of the economic boom of that decade. In this period, "capital investment in the automobile and all its spinoffs (including suburbs), though largely unanticipated and unplanned, rescued the U.S. economy from a period of growing stagnation and underwrote the economic boom of the 1920s" (Ashton, 1984, p. 64; see also Baran & Sweezey, 1966). However, with the general economic collapse during the Great Depression, suburbanization slowed significantly. In the attempt to revive the national economy, federal legislation was enacted that later would have significant impact on suburban development. The main thrust of this legislation was to create subsidies for owner-occupied, single-family detached housing.

In the early 1930s, new public agencies such as the Federal Savings and Loan Insurance Corporation and the Federal Housing Administration were created. These agencies provided for the first time federal guarantees on deposits in savings and loan associations (an important source for home investments) and on mortgages for newly constructed homes, thus encouraging the exodus of residential population to suburban areas. In addition, from the late 1930s the federal government engaged in a major project of war plant construction, which contributed further stimulus to suburban development. Data on public investment in wartime industrial facilities indicate that the federal government spent fully $17 billion on manufacturing plants and equipment during World War II, amounting to almost 43% of the total construction costs of all manufacturing facilities before the war (U.S. Surplus Property Administration, 1945, p. 18). Much of this industrial plant appears to have been built on remaining vacant land near the peripheries of central cities. This meant that city industrial workers moving to stay near their jobs often tended to relocate near the outer rim of the city and in its adjoining suburbs.

The full effect of these measures was not felt until the end of World War II because the Depression and war years held back housing investment. With the freeing of resources for investment in housing and highway construction after 1945, postwar suburbanization accelerated rapidly during the next two decades. The impact on central cities was devastating. Almost immediately following the war central cities began experiencing very slow growth or outright decline in employment and population, with manufacturing leading the decline into the 1960s. Already between 1947 and 1954 the annual rate of growth of manufacturing in the suburbs was almost a dozen times greater than in the central cities (Kain, 1970, p. 19). As of the mid-1950s similar trends could be observed in retailing and wholesaling as suburban shopping malls and warehouses convenient to freeways increasingly took the place of urban department store and warehouse facilities. By the early 1960s the central cities had registered sizable losses in total employment as suburban employment opportunities continued to expand rapidly: "Between 1954 and 1963, in the 24 metropolitan areas with populations greater than 1 million, the central cities lost more than 500,000 jobs while the suburbs were gaining over 1.5 million" (Ashton, 1984, p. 65).

During the 1960s the economic decline of the central cities in most established sectors of economic activity became quite clear. In traditional sectors concerned with the production of tangible goods such as construction and manufacturing, there were very sharp declines during this decade. There were also severe declines in traditional non-goods-producing sectors such as commercial trade, personal services, and transportation, communication, and utilities. Most of these involved blue-collar or gray-collar jobs. At the same time, there was a pronounced increase in employment in white-collar sectors ranging from public employment to professional and business services. Nevertheless, even in these sectors, suburban employment grew significantly faster than in the central cities.

Despite the increase in white-collar employment, by the 1970s major central cities had experienced serious losses in employment. In addition, the rapid continuing decentralization of urban households took its toll. In 1950 a little over one in four Americans lived in suburbia; by 1970 this ratio had grown to almost two in five. For the first time in America's history suburban residents represented a clear plurality of the nation's population. Suburbia had become the dominant settlement pattern, constituting nearly 38% of the total population, while central cities now accounted for only 31% of the total (Holleb, 1975, p. 12).

This net loss of population was only one of the factors in the weakening of the fiscal base of older central cities, which owed as much to the difference in economic status between largely middle-income out-migrants and lower income in-migrants as to the difference in their numbers. Due to the relatively more affluent status of the out-migrants, it is estimated that the aggregate annual income lost to central cities as of the early 1970s amounted to the difference between $55 billion and $26 billion, or a loss of almost $30 billion in annual urban household income due to migration (Sternlieb & Hughes, 1975, p. 9). Together with the postwar loss of jobs, the net loss of taxpaying household income constituted a major blow to the local economy and to the fiscal condition of the emerging corporate city.

From this brief review in marxist perspective of the decentralizing corporate city, it should be clear that market forces alone do not account for the rapid postwar growth of suburbia, which has had such serious consequences for central cities; corporate decisions to disperse industrial plants outside the cities and government policies to guarantee mortgages on new housing and to subsidize suburban highway construction were major factors in establishing the context for large-scale suburbanization. Nor is it likely that the operation of market forces alone will ameliorate significantly the fiscal problems of the corporate city that have followed from these developments; the policies of governmental and business institutions on issues ranging from taxation to programs for urban employment will have a significant role to play here as well.

On Government's Role:
Variety in the Marxist Perspective

As the marxists observe, the era of the corporate city coincides with the monopoly stage of capitalist development. In this period, the monopoly sector of the economy, composed of the largest corporations, is able to use government ("the state") to absorb certain costs that are intrinsic to urban development, such as the costs of urban infrastructure (roads and transport services, sewers, water systems) and the costs of educating and keeping healthy the urban labor force. No one individual firm could afford to pay for these facilities and services, nor does only one firm use them. The monopoly sector is able to share the costs of providing for these services by having the state define them as a public responsibility so that they are paid for out of government tax revenues, thereby distributing their cost far more widely than if individual firms tried to provide them. This has important implications for the role of the state in the shaping of urban development.

In general, marxists today would argue that the activities of the modern state indicate that it plays an important role in relation to the needs of corporate capital, in various aspects of urban development. However, contemporary marxism is not monolithic but spans a conceptual spectrum stretching through several different interpretations of the relationship between state and capital that differ over the role of the state and, in particular, the degree to which its policies are determined by forces external or internal to it.

Those marxists who are known as *instrumentalists* take a relatively limited view of the state, one that comes close to the classic marxist position that the state is essentially an executive committee for managing the policy interests of the capitalist class (Marx & Engels, 1967). These theorists view government as an instrument of the capitalist ruling class, or at least of the dominant fraction of that class, which uses government as a means of enforcing its own interests in capital accumulation. As evidence, instrumentalists emphasize the predominantly upper-class social background of policy-making personnel holding key government posts. This view can be criticized as reflecting a relatively simplistic external determinism in which the most powerful segments of the capitalist class directly determine the policies of government. Typically, such a perspective fails to note that modern government must be prepared to broaden its policies beyond simply supporting the capitalist accumulation process if it is to avoid having that process disrupted by social instability.

In contrast, the *structural* marxist view of the state sees the relationship between the capitalist class and the state as one that is "determined by the systemic constraints and contradictions" of the capitalist economic system (Esping-Andersen, Friedland, & Wright, 1976, p. 188). That relationship is determined in accordance with basic requirements imposed by the structure of the system, whether or not members of the capitalist class themselves participate in the institutions of the state (Poulantzas, 1972).[4]

If the economic system is to thrive, the institutions of the state must satisfy the system's requirements for political stability as well as economic growth, regardless of who staffs those institutions. In short, the relation between the state and the socioeconomic structure is taken to be an objective one; it does not depend on the specific subjective interests or the extent of governmental participation of members of the ruling class. In this perspective, the distinctive feature of the state is that it is the one institution in society that is able to act on behalf of the general interests of capital by transcending particular subjective differences of interest existing within the capitalist class. The state is viewed as having sufficient independence relative to any specific capitalist interests— *relative autonomy* in neomarxist terms—to act, if necessary, against particular segments of capital (such as crooked savings and loan executives and publicly reckless industrial polluters, for example) in order to preserve and legitimate the class structure of capital-dominated society as a whole.

This action includes the adoption of moderately reformist social welfare measures that smooth the frictions of class relations without eliminating underlying class differences. The structuralists regard the pursuit of welfare policies as an essential aspect of the role of the modern state, enabling it to cope with the systemic imperatives of advanced capitalist society. Contemporary social programs, including various urban programs, are seen as a structurally determined response to the requirements of advanced capitalism for social stability—the orderly social context necessary to the undisrupted continuation of economic processes of capitalist accumulation.

As opposed to instrumentalism, which is basically an extension of traditional marxism, the structuralist perspective represents an important step in the development of contemporary or neomarxist analyses of public policy development. Structural neomarxism articulates a complex theoretical framework for examining the specific roles played by government in the development of contemporary public policy. In addition, by advancing the concept of relative state autonomy even in a limited way, it has opened up the question of the degree to which governmental structures and processes may themselves influence urban policy independently of external pressure from urban policy interests.

The next section of this chapter examines how structuralism has been applied to the issue of the urban fiscal crisis. Following that, we will review some examples of another neomarxist perspective, that of class struggle or *class conflict,* which also relates to the concept of relative autonomy of the state. In the class conflict perspective, state policy develops as a response not simply to capitalist interests (whether their influence is assumed to be personal or structural) but to the prevailing balance of political power established in the ongoing conflict between the interests of capital and those of labor (Esping-Andersen, Friedland, & Wright, 1976, pp. 190-192). This perspective broadens the basis and complicates the dynamics through which state policy is decided. In general, it enables urbanists to analyze the conditions for development of urban and other social policies somewhat differently than the structuralist perspective by viewing class struggle as creating conditions conducive to state autonomy. Under conditions of well-organized working-class politics, the influence of labor may grow enough to counter selectively that of capital, opening opportunities for state managers to make policies free from simple determination by dominant capitalist interests. In this perspective then, sustained class conflict can provide the state with both the motivation and the necessary autonomy to pursue a program of social reform, even if that requires some reduction of support to capital accumulation and even if it raises the possibility of more fundamental challenge to the organization of the social system.

This is an important contemporary variant of neomarxist theory. However, it has some significant limitations for the analysis of urban programs in which interests that may not be primarily class related, such as those associated with race and ethnicity, act as sources of pressure on the formation of policy, as in the case of the complex multi-interest programs associated with the War on Poverty, for example.

THE CORPORATE CITY AND FISCAL CRISIS

As indicated earlier, the emergence of the highly concentrated corporate city was initiated by the rapid spurt of office skyscraper construction during the early decades of the 20th century. These developments were slowed and then virtually halted during the next two decades. In the 1930s, the Depression put severe restraints on central city development, only to be followed by the war years of the early 1940s, dominated by military priorities. By the 1950s central cities were beginning to feel the impact of two decades of deferred spending for maintenance or replacement of physical plant. At the same time, they were affected seriously by the accelerating out-migration of industry, commerce, and taxpaying population to the suburbs or beyond. In these circumstances, the supporting functions of the state took on key importance for the revitalization of central cities physically and economically. Business interests seeking to foster urban renewal in central city downtown areas did not rely solely on their own powers in the market. Increasingly they turned to government "to facilitate land-use changes . . . which were often difficult to achieve through costly haggling over patchwork land assembly. They also used the powers of government to construct the complementary public infrastructure that made their investments profitable" (Friedland, 1982, p. 78).

In the era of the corporate city, it is clear that government assumes an expanding role in both the national economy and the local community. Structuralist marxists argue that a direct connection exists between contemporary urban development and the growth of governmental spending. They view government expenditures as having two major functions: first, as aids to increasing productivity and hence capital accumulation; and second, as supports to the maintenance of urban social peace and public order. Government spending to assist directly productive activities takes the form of various public works and can be understood as *social investment* in support of private production. Such investment is aimed at providing physical capital in the form of infrastructure that facilitates or increases economic productivity. This ranges from basic utilities such as water, sewage, and electrical systems to transportation improvements, industrial parks, and research and development facilities.

Public spending to assist consumption takes the form of a variety of social service programs, which are referred to as *social consumption* expenditures. These are outlays that contribute to the maintenance and possible upgrading of the labor force, such as expenditures for education, health services, and housing. Such outlays contribute to creating a more skilled, healthy, and secure workforce, hence indirectly contribute to increased productivity. Together, social investment and consumption expenditures are public expenditures that assist the accumulation of capital by private enterprise. Thus they "socialize" many of the costs of private accumulation and hence are known as *social capital* expenditures. Finally, expenses for maintaining social order are referred to as *social expenses* and include expenditures for welfare and police-related functions (O'Connor, 1973).

In brief, in the era of the corporate city, which is the era of the dominance of large-scale corporate capital, government acts to provide physical infrastructure and social services that increase the productivity, lower the costs, and raise the profits of corporate business. This is the *accumulation function* of the state. At the same time, the state must maintain social peace and damp down social and political discontent. In part, it accomplishes this through social expense programs such as welfare and police activities, but these do nothing to improve work skills or productivity.

The state can accomplish these several objectives better through social service programs that contribute both to capital accumulation by raising productivity (e.g., through improving the skills or health of the labor force) and to the maintenance of social peace. One of the strongest arguments in favor of job training programs, for example, has been that they will help their clients become more productive and contented members of the community. To the extent that social service programs (funded by social consumption expenditures) contribute to reducing discontent and enhancing public order, they encourage sentiments that support the existing social structure as one that is appropriate and legitimate. Such program expenditures thus are viewed by structuralist marxists as essential to the *legitimization function* of the state (O'Connor, 1973).

As structuralists such as O'Connor point out, there is, however, a hitch in these arrangements. Ironically, the more the state assists the growth of private industry through its social capital expenditures, the more it lays the basis for further increasing both these expenditures and those labeled social expenses. As corporate industry becomes more efficient, its profits increase, and it can afford to invest in new, advanced-technology plant and equipment that increase production per work-hour, reducing the need for labor. Thus unemployment will tend to increase. Also, as physical production expands, it may enlarge the volume of pollutants in the environment, increasing the need for antipollution and public health expenditures. And the more industry goes high-tech, the more the need to increase expenditures for educating the workforce. This in turn contributes to the increased cost of social consumption expenditures. Finally, if high unemployment due to technological displacement becomes persistent, there will be increased social expense outlays for welfare and police services.

A basic theme of the structuralists is that despite outward appearances, modern urban development is not essentially the product of a cumulative process of technological evolution. Rather it is the result of changes in the mode of economic production and in the role of the state, which interact to produce a distinctive new system of capital accumulation and of social control of the labor force. Transitions between phases in the development of this new system typically occur in response to crises based in fundamental internal dilemmas or contradictions in the operation of the system that arise during a given historical epoch.

The fiscal crisis is the result of one such set of contradictions; in particular, it is a consequence of the fact that public subsidy for private capital accumulation often tends to increase not only productivity but also its related social and environmental costs. Thus the very success of government support of corporate business leads to increasing government outlays. Once these expenditures reach a certain point, however, they begin to compete for budgetary resources with the original expenditures that were aimed at increasing business productivity and profits. Reduction of those outlays tends to slow down the economic growth out of which must come the revenues to pay for both compensatory programs and further subsidies to business.

Overall, the fiscal crisis of the state results from a combination of "increased demands on the state . . . and of the inability of the state to expand sources of revenue fast enough to meet increased demands" (Gold, 1975, p. 130). As noted earlier, in the era of the corporate city government assumes a dual responsibility of keeping the economy growing and maintaining social peace. Because government does this in the context of a capitalist economy, it also incorporates the class contradictions that are intrinsic to that economy. In particular, it accepts the fact that although the social costs of production such as unemployment and pollution are generated privately, they will be paid for publicly under the elaborate system of government subsidies that develops in this period. In addition, it accepts the fact that although these costs will be paid for publicly, the profits that flow from the production process are collected privately. This makes it problematic that the revenues to cover these costs either will be distributed equitably or will be adequately available. As a consequence, "fiscal crisis—the gap between expenditure demands and available revenues—becomes the state budgetary expression of class conflict in a monopoly capitalist society" (Hill, 1984, p. 301).

In the United States the major local sources for funding municipal government activities are taxation and borrowing, together with various charges and fees for public services and utilities. In the past two decades, government borrowing through the sale of bonds has grown as a source of financing public expenditures. However, when interest rates have been high (as they were over much of that period), government has had to raise the rates it pays in order to attract buyers to its bonds. This has raised government's costs of borrowing, which are paid out of budget revenues, and has thus contributed to fiscal stress and limited the ability of local governments to continue financing their spending through public borrowing. At the same time the political difficulty of raising tax rates or increasing tax assessments has grown considerably, given rising public sentiments for "no new taxes." Further, contemporary marxists emphasize, taxation has become limited as a means to relieve fiscal stress also because under prevailing tax laws, profits from investment have become significantly exempt from taxation (Gold, 1975).

Moreover, compounding the problem at the local level is the fragmented system of government that is typical of metropolitan areas. As Hill (1984) observed, "Uneven economic development in a politically fragmented metropolis fosters uneven fiscal development among local governments" (p. 302). As major investments shift to the outer suburbs or out of the industrial heartland altogether to growing metropolitan centers in the South and West, unemployment and poverty become concentrated in older central cities and the older inner-ring suburban areas that surround them. In this way, the tax base becomes separated from social needs, and older cities find themselves unable to raise the revenues necessary to meet rising expenditure demands.

In addition, in the large metropolitan areas, central city social expenditures tend to multiply with the increasing size and complexity of the area. For example, central city investments in education tend to benefit other jurisdictions when the city's middle class and those with specialized skills migrate to the suburbs and to other metropolitan areas. Central city expenditures on transportation, hospitals, and cultural facilities benefit the region as a whole, but the local tax bill is often paid by central city residents alone. Aging central cities also confront added demands for social expenses as displaced people spill in from declining rural areas and as part of the central city labor force is rendered redundant by technological change. The burden of these expenditures falls heavily on central city property owners and workers, although the major benefits go to the suburban fringe and the nation as a whole (Hill, 1984).

In sum, capital accumulation and urbanization require expanded social investment, social consumption, and social expense outlays by local governments. The geographical movement of capital according to the criterion of profitability, within an elaborately divided state structure characterized both by federalism and by a fragmented system of local governments, means the divorce of the tax base from social needs and expenditure requirements and a tendency toward fiscal crisis in older urban centers in the United States.

URBAN CONFLICT AND THE MARXIST PERSPECTIVE

For marxists, the emergence of the fiscal issue is the visible face of underlying class conflicts played out in the fiscal arena. During the Reagan administration this appears to have become quite clear. Tax cuts, together with other facets of tax policy, were aimed at placing capital resources in the hands of the well-to-do, under the theory that this would stimulate useful investment in economic growth. On the other hand, to pay for these tax cuts, net expenditure cuts (counting inflation) were made in social services for low-income workers as well as in income maintenance programs for the nonworking poor. Various social critics referred to such Reagan administration policies as examples of a new class war waged on behalf of the wealthy against America's working class and poor (Piven & Cloward, 1982).

In connection with this theme, among marxist urbanists the interregional and international mobility of corporate capital—the footloose quality of contemporary corporate investment that these urbanists view as having contributed significantly to the urban fiscal crisis—has been seen also as an expression of a conflict of class interests. As Tabb (1978) observed:

> Economic and political struggles, not merely some sort of technological or climatic imperative, explain both the pull of the Sunbelt and the push from the Northeast. . . . The attempt of workers to better their conditions, and of capital to improve its profits by cutting costs, has, as one of its aspects, geographic mobility—a weapon workers are finding their employers more and more willing to use. (p. 249)

Tabb spoke here of the interests of workers as against the interests of capital as if these were two internally undifferentiated blocs. The tendency to do so, particularly with reference to workers or the working class, is apparent also in the work of David Harvey, who sometimes has been referred to as the leading marxist urban theorist. As Harvey (1978) wrote:

> The domination of capital over labor is basic to the capitalist mode of production; without it, after all, surplus value could not be extracted and accumulation would disappear. All kinds of consequences derive from this, and the relation between labor and the built environment can be understood only in terms of it. (p. 10)

Harvey observed that through its particular organization of the work process, industrial capitalism separated the workplace from the place of residence. This in turn created two spheres of class conflict: one in the workplace over wages and working conditions; the other, in the place of residence, concerning costs and living conditions, conducted against secondary forms of exploitation represented by merchants, landlords, real estate agents, and the like.

Harvey (1978) envisioned a number of different scenarios in the struggle of workers to defend their living standards. In one, each worker pursues his or her own interests and tries to command for private use the "best bundle of resources in the best location." This would involve a "competitive war of all against all" based on the ethic of "possessive individualism." Another situation involves community consciousness and community action of some type. Home-owning workers will seek to protect the value of their property from negative externalities in the local community and to ensure high standards of public service. One facet of such community consciousness may be competition among communities for government protection or investment. This situation may lead to "internecine conflicts within the working class along parochialist, community based lines" (pp. 31-33).

The last situation that Harvey (1978) described is that of a "fully class-conscious proletariat struggling against all forms of exploitation, whether they be in the work place or in the living place." Here all workers struggle collectively to improve their common lot, and "eschew those parochialist forms of community action which typically lead one faction of labor to benefit at the expense of another (usually the poor and underprivileged)." These three situations then— possessive individualism, parochial communalism, and proletarian collective consciousness—are "points on a continuum of possibilities." Although Harvey conceded that in the United States labor appears to be "strongly dominated by competitive individualism and community consciousness," the implication remains that labor simply cannot be taken for granted and assumed to be located at any particular point on this continuum. The chance of attaining working-class consciousness is always there (pp. 31-33).

Although Harvey's (1978) discussion of a *fully class-conscious proletariat* is rather idealized, he does recognize that this is only one of several possibilities, particularly in the American context. Nevertheless, the general picture he provides is nowhere as complex as the prevailing reality of the contemporary American urban scene. The American city cannot be divided validly between capital (in its industrial, financial, or mercantile manifestations) and the various other segments of the community, lumped together as "labor." It is not merely that labor is divided into various factions or subclasses, based on differences in race and ethnicity as well as on level of occupational skill and of union organization, but that much of the community's labor force does not view itself as labor, that is, as working class. This is true particularly of members of the new white-collar workforce who insist on seeing themselves as middle class, hence resist unionization or other actions that might link them with the blue-collar workforce precisely because the latter are perceived as a traditional proletariat.

Thus, although Harvey (1978) described the condition of the fully class-conscious proletariat as a point on a continuum of possibilities, the possibility of achieving this condition, which may well have existed during the era of the industrial city, now is diminished seriously in the day of the young upwardly mobile urban professionals. Speaking in terms of classic marxist theory, educated white-collar professionals who increasingly can be found working on salary in bureaucratic organizations—ranging from engineering or law firms to universities and health maintenance organizations—validly may be considered a "new working class" because they own neither the means of production nor the means of administration of their work process. Nonetheless, it is quite unlikely that many of them will be ready in practice to identify with such a classification or that they would be willing to act on it in accordance with "proletarian class consciousness" (Ehrenreich & Ehrenreich, 1979, p. 17). Thus, in American cities today, there appears to be lacking a class-conscious proletariat ready to struggle in a determined and united way against the dominance of capital in the shaping of the urban future. This circumstance raises serious problems for marxists in deciding how to move from critical interpretations of urban change to practical strategies for shaping urban development.

URBAN DEVELOPMENT IN MARXIST PERSPECTIVE

In further considering marxist responses to contemporary urban development, we will examine several "trajectories of urban change" plotted by Hill (1984) in an imaginative chapter on fiscal crisis and urban policy alternatives. Each trajectory is a possible scenario, an image of a possible future for the troubled central city. Any particular city may exhibit each scenario in varying degrees.

The first image is that of the *pariah city* (Hill, 1984). This is "a form of geographical and political apartheid"—a "reservation for the economically disenfranchised labor force in a monopoly capitalist society" (p. 311; see also Long, 1971). Such a city incorporates "the poor, the deviant, the unwanted and those who make a business or career of managing them for the rest of society" (Sternlieb, 1971, p. 16). The pariah city is a welfare reservation, in which a combination of social service professionals and police make the containment of the natives simply a business. Here the poor and unemployed serve as conduits for the transfer of state revenues to inner-city slumlords, merchants, loan sharks, and service professionals who in turn funnel the surplus outside of the community. There is no way for residents to gain control of the situation; capital outflow and the deteriorating tax base make central city leaders ever more dependent on state and federal funds while White-dominated legislatures remain indifferent to the underlying problems of the pariah city. As a consequence, despair and apathy define its basic mood.

As the congressional Joint Economic Committee (1982) observed in a fiscal survey of 48 large cities:

> A city which is in the process of raising taxes and cutting services, and which is resting on a decaying infrastructure, provides no inducement to business expan-

sion or in-migration. .... As the city service levels and physical plant continue to deteriorate in conjunction with increased costs to residents and businesses, those that can will heed the President's advice and "vote with their feet." (pp. viii-ix)

Despite these bleak prognoses, Hill (1984) noted that a big city is not an isolated reservation. It typically stands at the core of a larger metropolitan system with which it maintains significant economic and social relations. There is simply too much at stake to abandon older big cities to stagnation. Despite their recent reverses, they remain important centers for their own metropolitan areas, as well as control centers for clusters of governmental, economic, and cultural institutions that extend far beyond these cities themselves. In short, they represent huge economic and social investments that simply cannot be left to decay.

The favored solution offered by the corporate establishment to the urban fiscal crisis involves development of the *corporatist city* (Hill, 1984). This amounts essentially to the corporate city which was described earlier, with two important added dimensions: Its government would embrace its entire metropolitan region, and it would be run strictly according to modern business principles, as if it were a corporation. Much in this conception of running the central city as a big business resembles basic notions of the early 20th-century urban governmental reform movement discussed in the next chapter, except that this time the city is projected to full metropolitan scale.

The corporatist scheme for governance of such a city would link governmental structures at the state, metropolitan, and local levels into a type of supraurban political system managed according to hierarchic principles of corporate planning, in effect, joining these governments into a new vertical-interorganizational corporatist network. This in turn would require basic changes in revenue collection and in the production and distribution of public services in the metropolitan area. In such a governance system the collection of taxes and the control of public service production shifts to higher levels than the traditional central city government. The relatively inelastic property tax is replaced by a combination of income, sales, and value-added taxes, which are administered by metropolitan or state government to ensure the capture of suburban revenues. Centralized administration and budgetary planning provide organizational and technical means by which budget priorities can be set to support corporate capital and enhance its profitability. As corporate earnings increase, government tax revenues follow. Increased revenues are used to subsidize corporatist solutions to a wide variety of urban problems. As a result, social tensions in the central city are reduced through social programs such as the Model Cities program, which offer a combination of improvements in the local community, the upgrading of individual skills, and the provision of expanded job opportunities.

The key to the corporatist solution to current urban problems is the willingness to depart from traditionally limited governmental jurisdictions in order to provide a rationalized system of at least metropolitanwide governance. However, the jurisdictional changes involved carry with them some serious political problems. For one, the transfer of fiscal and administrative responsibilities to state or metropolitan governments poses a clear threat to the privileged status of suburban areas, enshrined in law and tradition. This is probably the major reason why most attempts at metropolitan government have failed. In addition to threatening influential suburban professional and political interests, the corporatist city also implies rising taxes for the urban workforce and for small business located in the city, beyond what they might consider reasonable in return for the possible benefits.

Finally, "scientific management" of urban government involves the likelihood that a corporate agenda would predominate in rationalizing city budgeting, as well as in reducing the prerogatives and wages of public employees. Consequently, the chances of urban fiscal problems being ameliorated by corporatist urban reforms depend in some measure on whether new political support can be found—possibly among growing urban minority populations that have generally cool relations with organized labor and a real stake in occupational and community improvement—to counter potential sources of political resistance both from within the city and from its suburbs.

Hill's (1984) final urban scenario involved his conception of the *socialist city*. For him, such a city represented in the long run a model for a fundamentally restructured political economy. As he noted, its establishment would require substantial government investment in the local economy, together with the fostering of a more democratic planning process. We will review first Hill's concept of the realized socialist city and then his examination of some of the obstacles to its realization. Particularly in detailing specific strategic problems that might be involved in the transition to a socialist city, his discussion provides an example of some of the possibilities of the class conflict perspective and its usefulness as an analytic approach.

In Hill's long-term model of the socialist city, the community owns the community's wealth, so that whatever economic surplus is produced goes back to the residents rather than to bankers, landlords, or real estate speculators. By channeling surplus wealth created by local economic activity directly to the city treasury through taxation or other means, the city helps insure itself an adequate resource base with which to improve and expand public services and facilities. The larger society does not have to be already socialist to begin the transition from the contemporary capitalist city to the socialist city. Municipal ownership of utilities or other enterprises makes it possible to generate revenues through user charges. In addition, the city can acquire property by use of its powers of eminent domain through foreclosures on properties that are seriously tax delinquent or through municipal purchase. It then can lease the property to private business at a rate that allows an economic return to the city, thereby contributing to the city's social surplus. City employees' pension funds can be invested in profit-making community ventures, except that the profits would accrue to the city treasury rather than to private investors. City funds accumulated through these various avenues then could be used to provide loans to local community groups organized as economic cooperatives to provide improved housing, food supplies, or credit to their members. Jobs would be created in these cooperatives for local citizens, money would be retained in the community—instead of flowing to private businesses based outside it—and local economic development thereby would be encouraged.

In economic terms, the socialist city implies "a major redistribution of wealth," and "massive investment in public sector activities and employment" (Hill, 1984, p. 316). In Hill's perspective it also implies a new approach to decision making for the community's economy. The market is subordinated to a democratic planning process that includes citywide hearings that provide for active representation of autonomous community groups. Presumably receiving public subsidy, these groups are well staffed and funded and have access to independent sources of information. Besides the enhancement of local democratic participation, there is an economic payoff; with experience, community groups can develop familiarity with appropriate small-scale technologies that are resource conserving and can help meet basic needs in areas such as local energy provision and transportation. Overall, urban development planning changes from a technocratic activity conducted by city planning experts to a process that is explicitly political as well as technical, incorporating democratic bargaining between representatives of business, labor, local communities, and the city government.

Clearly aware of the difficulties of implementing such a scenario, Hill (1984) concluded by reviewing some of the major barriers to the creation of the socialist city. The political and economic realities of life in capitalist society tend to place serious constraints on redistributive public programs and on public sector production of goods and services that seriously may threaten private profit. Moreover, the various elements of urban labor are not necessarily politically united around a common agenda for future urban development. As he noted, there are divisions between city employees, community organizations, and private sector workers conscious of their own position as taxpayers.

What gives him hope is first what he refers to as the "socialist undercurrent" in local politics—the emergence of a populist strain based on the various grassroots movements, from civil rights to the welfare and consumers' movements—that has surfaced since the late 1950s. This has helped elect political officials who have called attention to the need for strengthening public control or extending public ownership over some forms of private capital, such as banking, energy utilities, and insurance companies, and who have tried to include this on the governmental agenda along with proposals for increased citizen participation and government accountability.

In addition, Hill believes that divisions between municipal employees, private sector workers, and community groups are not necessarily permanent. Under conditions of urban fiscal crisis, municipal employees are faced increasingly with the prospect of wage and benefit cutbacks or loss of jobs. At the same time, Blacks and other minority communities are a growing part of both the municipal labor force and the urban poor and have gained a certain degree of political leverage due to their numbers and organization, particularly in older big cities. This suggested to Hill the possibility of coalitions between public workers and local community activists for the expansion of public services, particularly to the neediest, and the upgrading of public jobs. However, his optimism is more guarded with respect to relations between municipal workers and private sector labor, precisely because he is more realistic than Harvey in recognizing the existence of a "fragmented urban working class." One key problem is that municipal union demands for improvements in wages or working conditions frequently mean increased city taxes to a central city working class that includes many nonmunicipal workers who are already highly taxed. Hill's (1984) response to this problem was carefully conditional. In his terms:

> Union demands . . . will at some point have to be tied to changes in methods of revenue accumulation and overall city budget priorities. To the extent that city workers demand that their material needs and conditions of work be met by reallocating available resources rather than by increasing taxes, and . . . that they challenge the priorities of corporatist reforms, present divisions between private and state workers and neighborhood organizations will narrow. (p. 319)

Apparently what is meant by "changes in methods of revenue accumulation" is revision of the local tax structure that regressively weighs most heavily on lower income groups. Just what changes in budget priorities this would involve are not elaborated. To the extent that such changes might allow increased spending on social programs for the poor, this probably would win support for the suggested reforms from residents of inner-city neighborhoods and from public sector workers associated with those programs, only to lose it from private sector workers. This merely highlights the dilemma of urban radicals in seeking the unified support of the urban working class, whose internal divisions reveal multiple lines of cleavage.

One way to overcome at least partially these strategic obstacles is to go beyond the local political system to higher levels of government, a lesson learned by urban liberals in their earlier pursuit of antipoverty programs. Thus Hill (1984) is quite correct in observing the need to link local politics and its limited revenue bases to "vertically integrated organizations influential in state capitols and in Washington" (p. 318). We will explore the implications of this observation more fully in the next chapter in the discussion of intergovernmental relations. Through such organizations access may be gained to valuable political and economic resources required at the local level. Thus a full understanding of urban development processes and problems requires a sense of the organizational and interorganizational dimensions of urban policy formation, to which we now turn.

NOTES

1. For a sampling of work reflecting variations on the marxist perspective as applied to issues of urban development theory, urban policy, politics, and social movements, see Fainstein et al. (1983); Smith (1988); and Tabb and Sawers (1984). For the views of exponents of a nonmarxist urban conflict perspective relating to postwar movements for neighborhood government or community control, see Altshuler (1970) and Kotler (1969).

2. For a discussion of the role of corporate institutions in the shaping of urban change since the mid-19th century, see Gordon (1984). See Friedland (1982) for a focus on the corporate role in post-World War II urban renewal, including interlocking relations between large corporations and commercial banks in carrying out urban development and corporate influence over the shaping and implementation of renewal policy.

3. Kain's (1970) work on the forces shaping American metropolitan development in the first two decades following World War II emphasized the central importance of locational decisions made by industrial corporate elites, rather than the consumer preferences of individual households. As he observed: "The location of manufacturing is especially critical in determining metropolitan spatial structure, since the locational decisions of most manufacturing firms are largely unaffected by the distribution of metropolitan population. Manufacturing determines the location of urban households, not vice versa" (p. 20). For discussion, see Solomon (1980).

4. As Poulantzas (1972) observed, "If the function of the state . . . and the interests of the dominant class coincide, it is by reason of the system itself; the direct participation of members of the ruling class in the State apparatus is not the cause but the effect" (p. 245).

[3]

Introducing an
Interorganizational/Policy Perspective

To this point we have examined the ecological and marxist perspectives on urban development. Despite important differences each is essentially an economically oriented theory of urban development. Contemporary ecological theory is largely a theory of the developing space-economy of modern urbanism as urban development moves beyond the boundaries of the industrial city and beyond industrial era definitions of urbanism. This theory focuses on the role of economic and technological factors in the spatial expansion of urbanism during the 20th century. The process of development is seen as natural and functional, one in which late-20th-century urban settlements adapt to postindustrial economic transition, suburban growth, and associated technological innovation through central city economic and physical restructuring. The appropriate role of government is seen as the development of policies that will support and ease the urban adaptation to economic transition and related spatial change.

Marxist theory also revolves about an economic core in that it views historical changes in the structure of the city as a consequence of changes in the society's economic mode of production. Accordingly, contemporary marxism is much concerned with investment decisions made by private capital—especially corporate capital—as it works to expand its scope and powers of accumulation with the assistance of government, whose programs on one hand support major private investments and on the other try to cushion their social impacts.

A major point of difference with contemporary ecology lies in the marxist emphasis on the objectives of large-scale corporate capital—aimed at control of the labor force and other factors of production—and the influence these objectives have on the decisions of capital, particularly with reference to the geographic location of corporate productive and administrative facilities. The marxists view the logic of capitalist locational decisions rather differently than the ecologists. They emphasize that such decisions are affected significantly by considerations of the social control of capital over labor and not just by technical considerations of locational effects on economic efficiency. At the same time, however, their relatively limited perspective of the function of government restricts it to being basically an auxiliary to capital in the shaping of the urban development process.

In short, despite their differences, contemporary ecology and marxism tend to converge toward a view of the role of public policy as supporting and reinforcing the decisions of capital. For the most part, neither theory focuses on the possibilities of a significant role for public policy. The role of policy is viewed as essentially subsidiary either to technoeconomic forces or to dominant economic class interests. Policy either supports these predominating factors or at most is moderately ameliorative, acting to partially compensate for their negative impacts or failures (as in the case of limited public programs to provide housing for the poor, which the private housing industry either cannot or will not provide). Neither theory examines in detail whether governmental policy can have a creative role, opening up opportunities for positive change not determined either by the market or by the capitalist elites that are seen to control urban development. A more comprehensive theory requires the full articulation of the role of politics and public policy in relation to the roles of economy and technology in shaping the specific direction that urban development takes.

BACKGROUND TO THE
INTERORGANIZATIONAL/POLICY PERSPECTIVE

In contemporary society, to speak of policy typically is to speak of organization, particularly large-scale organization. Yet the major relevant theories discussed so far tend to emphasize the economic dimension of urban development to the relative neglect not merely of its political-governmental dimension but also of its organizational dimension. In my perspective, the processes underlying contemporary urban development cannot be explained fully unless they are understood as expressions of a deep organizational transformation in postindustrial American society and its urban areas. Hence this transformation should be examined not only in its economic and political aspects but also in its organizational aspects. Large-scale organizations have assumed special functions and significance for contemporary American society and for urban and suburban communities within that society. As Turk (1977) observed, in many respects modern life has become organizational life. The fortunes of modern societies depend on a variety of organizations ranging from schools and churches to governments and armies, so that organizational functions and interorganizational relationships are essential to understanding contemporary society and its communities:

> The actions and interactions of organizations . . . form the affairs of cities, nations, and still larger social units—and even constitute their identities. . . . Interorganizational relations are everywhere in evidence. . . . Indeed . . . the attempt seems warranted to view [urban industrial or postindustrial society] . . . or any one of its major subdivisions as a patterned aggregate of organizations. (pp. 1-2)

In seeking to relate the organizational dimension to the economic and political, I draw on the theory of the *political economy of development* (Ilchman & Uphoff, 1969). In this perspective, society's productive assets include organization, information, and political authority as well as traditionally recognized economic factors such as land, labor, capital, and entrepreneurship. Viewed this way, the modern development process not only seeks to enhance the traditional factors of production; it also seeks to expand the availability and capacity of organizational, informational, and authoritative social decision-making factors at all levels of society, including the local community. Thus rationally purposive community development is seen as something more than just the spontaneous result of numerous uncoordinated transactions in the private economy. It cannot be left simply to the chance outcomes of the economic marketplace. Instead, rational development is conceived in terms of the conscious exercise of public policy in conjunction with significant economic interests active in the marketplace to achieve socially constructive outcomes that can enhance both productivity and community. If necessary, this may require change in the structures of public authority and in the practices of private management in order to improve capacities for rational planning and cooperative interaction within and between organizations engaged in development activities of joint public-private interest (Hanson, 1983).

The interorganizational/policy approach directs attention to the role of public-private organizational networks in linking communities to sources of information, economic resources, and political authority vital to enabling appropriate public action and stimulating relevant private participation in the urban development process. In this perspective, it becomes clear that the public sector has in recent decades taken the initiative in providing important support to urban development through vertical program delivery systems carrying resources from higher levels of government (Milward, 1982; Warren, 1978). On a national scale, this often has been accomplished through federally created urban programs and the interorganizational networks that connect them to related public or private organizations at the local community level. Examples include the Urban Renewal program, the Community Action and Model Cities programs, and the Community Development Block Grant program. These will be discussed in detail in later chapters.

In order to understand the national urban policy system in both its horizontal and vertical dimensions, it is helpful to refer to the concept of the *interorganizational network* developed by Benson (1975):

> The interorganizational network may be conceived as a political economy concerned with the distribution of two scarce resources, money and authority. Organizations, as participants in the political economy, pursue an adequate supply of resources. . . . The interorganizational network is itself linked to a larger environment consisting of authorities, legislative bodies, bureaus, and publics. The flow of resources into the network depends on developments in this larger environment. (p. 229)[1]

This conception succinctly defines the notion of the interorganizational network from a political economy perspective and is well suited to the analysis of urban development today. The modern city can be viewed in terms of a complex of relationships involving the exchange of political and economic resources among different urban organizations linked together in networks of interdependency. The interorganizational/policy perspective helps bring into focus the role and structure of organizational networks that act as linkages between different sectors of the developing city, particularly between organizations of the urban economy and those of its political system, as in the case of the political machines that acted as vital organizational networks of resource exchange in the governance and development of the early 20th-century city.

The interorganizational perspective also helps identify external linkages that during recent decades have served to connect the organizational systems of older cities to resources provided through larger regional, state, or national networks. It thus helps account for the emergence of the *intergovernmental city* of the later 20th century,[2] characterized by its high level of dependency on vertical resource linkages to federal or state government agencies and their programs. In addition, it makes clear the significance of relations between urban organizations and other organizations not physically located within the city, whether they are agencies of higher levels of government or private corporate headquarters organizations. Although remote from the local community, these policy-making organizations exert important influence on it through a variety of interorganizational resource relationships. Today, both the making of urban public policies and the shaping of the decisions of private firms occur within the context of such relationships, with consequences for numerous issues related to the future of urban centers—ranging from urban residential livability and the quality of local educational institutions to central city attractiveness to business location and the retention or migration of taxpaying residents and skilled labor force.

URBAN GOVERNMENT IN THE INDUSTRIAL CITY: POLITICAL MACHINES AS INTERMEDIARY ORGANIZATIONS

A basic objective of this book is to examine the changing roles of interrelated public and private organizations in setting the context and deciding the content of urban policy in contemporary America. In preparation for that, this section provides a historical perspective of governance in the early industrial city and of the role of political machines as organizational intermediaries between the public sector and selected elements of the private sector during that era. This should help to appreciate better some key issues underlying the later shift to bureaucratically organized urban government, particularly in older big cities.

Organizational reforms in urban government over the past half century have increased significantly the power of bureaucratic public agencies to affect the process of urban policy making. By contrast, in the industrial city of the late 19th century, control over public policy making tended to be concentrated in a stably dominant political party, rather than in the city's public agencies. Such parties came to be known as "political machines" due largely to their highly disciplined electoral organizations, which often achieved such precision that machine politicians knew in advance almost exactly how many votes they could count on in any given election district. The result was an exceptional ability to choose the city's public officials, which carried with it the power of decision over a wide range of public policy issues.

Political machines enjoyed increasing importance in the era of the developing industrial city, roughly from the mid-19th century to the early decades of the 20th century. This was a period of freewheeling development characterized by rapid urban economic and social change. America was still a nation dominated by ideas of limited government, in which the growth of the cities was regarded as a natural process flowing from the independent decisions of many different entrepreneurs operating in an expanding urban marketplace.

The typical form of urban government inherited from the early 19th century was that of a "weak mayor" linked to a strong legislative council that had authority to override the mayor's administrative decisions at many points. Although this political structure might have been adequate to the needs of a slowly moving preindustrial town, in the industrializing city it placed severe restraints on executive power and encouraged the members of the council to behave as if they were a political marketplace open to the highest bidders. As a consequence, the weak mayor lacked sufficient authority to control the affairs of the administration with regard to either budgetary matters or the appointment and removal of the heads of municipal departments. The result was a governmental structure that essentially was ineffective for providing central leadership in an ever more complex and problematic urban environment.

During this era of structurally weak local government, the political machine was able to step into the resulting gap in urban governance and play the role of a type of informal, spontaneous policy-making center. *Spontaneous* here implies a relative absence of planning either in particular decisions of the machine or in the overall pattern; policy making in the machine was ad hoc and piecemeal, dictated more by political expediency or chance of economic payoff than by consistent principles of good long-term planning. Nevertheless, the machine provided the central focus of public policy activity in the industrial city. Like no other organization, it assumed command in the making of authoritative decisions concerning the allocation of public resources in rapidly developing industrial centers.

The machine developed through several stages as an organizational adaptation to the fast-changing conditions of 19th-century American urban life and politics. In its operations, politics was treated largely as a business activity, engaged in for personal gain. As a political organization it spanned diverse urban interests that worked together in coalitions of practical convenience, rather than being bound by higher political principles. In its origins it was often the creation of political entrepreneurs of immigrant background who saw the continuing influx of large numbers of European ethnics as a rich but still unorganized source of political power. By the end of the century, machines in the larger cities generally had achieved the position of tightly disciplined political parties run on the basis of "inducements that are both specific and material," as well as inducements that were nonmaterial but no less important, such as political and social status for their members and supporters (Banfield & Wilson, 1963, pp. 115-117).

The machine's inducements were essentially invitations to its supporters to engage in relationships of exchange, trading off the public jobs, city contracts, legal favors, and political recognition that it controlled as payoffs in return for their votes, money, time, and energy. The machine's main constituencies could be grouped in two major blocs: the ethnic immigrant populations that typically occupied the political wards of the inner city's working-class districts and upwardly mobile businesspeople, located particularly in and about the central business district. In effect, the machine linked itself with two streams of development in the growing industrial city, the development process of the ethnic working-class communities and that of the rising segments of the business community.

The Machine and the Ethnic Communities

The traditional image of the machine that has been handed down histori-cally identifies it most closely with the urban ethnic immigrants. Thus Ban-field and Wilson (1963) described the machine's relation to the ethnic wards as a natural product of the industrial city's diversity:

> Where a city is made up of distinct natural areas or subcommunities, its politics often reflects these attachments and intensifies them. Ward boundaries are usu-ally consciously drawn to reflect the ethnic, religious, and class divisions within the city, and many wards are still highly homogeneous. ... In such places, politi-cal organization and ethnic organization are closely related ... and one is more or less created by the other. (p. 51)

Exchange relations with the machine carried advantages for individuals in the ethnic communities; if a breadwinner lost his job, a youngster ran afoul of the law, or a family was burned out of its apartment, the machine could be helpful in setting matters right through its connections with city hall and the munici-pal agencies. This aspect of the machine's relationship to its local constituents is the one that typically has been most emphasized. It exemplifies a type of *distributive policy* in which favors are dispensed separately, case by case, to individuals or families in accordance with the loyalty and intensity of their support.

At the same time, the ethnic community as a whole could derive various benefits from its connection with the machine. Through its many channels the city machine connected not only with individual supporters but also with the organizational complex of the ethnic community. A key linkage here was the local political clubhouse, whose active participants were drawn from a wide variety of voluntary groups, ranging from benevolent associations to fraternal lodges to the lay organizations of local churches. In this regard, the Catholic Church and its associated network of lay organizations played a central role in the political development of many ethnic communities. Generally, the parish organization of the church tended to bind Catholic ethnics together in geographic units where they lived, worked, worshiped, and sent their children to parochial school. This community centeredness proved to be of considerable assistance to the growth of local political participation and the emergence of well-organized voting blocs; over the long run, it also contributed to the ethnic community's economic development (Harrigan, 1985).

The turn-of-the-century Irish community serves as an example of how strong orientation of the faithful to the local parish supported the long-term growth of community facilities and the expansion of economic opportunities for various members of the community. The readiness of the local laity to donate private resources for investment in community institutions made possible the construction of a broad range of religious facilities. According to Harrigan (1985):

> Not only were churches and schools built, but so were rectories, convents, parish halls, cathedrals, chanceries, high schools, bishop's residences. . . . All of this construction activity generated a significant number of jobs and construction contracts . . . to Irish businessmen or at least to businessmen who were willing to hire Irish laborers. The early decision to create parochial schools was a built-in guarantee that some Irish could build successful businesses in the construction trades, trucking, real estate, insurance, and related business enterprises. (pp. 52-53)

The Irish community provides a model of a community whose strong central institutions acted to concentrate private family and business resources and target them to community projects to a degree that otherwise might not have been possible. Over several decades this contributed to the development of the community not only physically but also economically and socially, providing added employment for the working class and further opportunities for the growth of a business and professional middle class. In this process the church acted as an organizational hub of the community, located at the core of an interorganizational network consisting of families, businesses, and voluntary organizations. Overall, this was an important internal developmental network whose contribution to local economic growth was essentially to mobilize and concentrate otherwise unutilized local resources for neighborhood development.

In addition, for those neighborhood communities that also had strong ties to the political machine, the process of internally generated development received a considerable boost from outside the community. In those cases, self-initiated internal community development was amplified and accelerated by the flow of public resources into the community in the form of public works, contracts, and services in return for local contributions to the machine. Thus the machine augmented local economic development by acting as an intermediary organization that linked the resources of city government to the developmental process in supportive local neighborhoods. In short, the machine reinforced the development process through its pursuit at the neighborhood level of ad hoc distributive policy, which although typically piecemeal and unplanned made a contribution to economic growth in selected local areas. As discussed next, the machine assumed a similar role with respect to the urban business sector on a substantially larger scale.

The Machine and the Urban Business Community

Besides the ethnic neighborhood community, another important segment of the growing industrial city with which the machine developed important organizational connections was the central business sector. The machine typically became involved in the distribution of public franchises and contracts to growing business firms, together with subsidies in the form of the provision of a wide variety of infrastructure from street paving and lighting to the running of trolley lines in the vicinity of growing industrial and commercial districts. Business owners found that payoffs to machine functionaries facilitated access to police protection and fire-fighting services, as well as water and sewage hookups and other essential services, all of which would have been extremely difficult to organize and provide privately, not to speak of the expense that was spared private entrepreneurs who instead were able to share it with the general taxpaying public.

This is what marxists refer to as the socialization of the infrastructure expenses of urban private enterprise. Many hundreds of millions of dollars flowed into the provision of infrastructure to serve business districts, constituting a huge public subsidy to continued economic growth in the urban business sector (Glaab & Brown, 1967). The achievement of the machine politicians was that by acting as political brokers, they were able to establish exchange relationships with urban business elites at the same time that they maintained such relationships with the communities of the ethnic working class. As Scott (1969) has observed:

> Frequently, a three-cornered relationship developed in which the machine politician could be viewed as a broker, who, in return for financial assistance from wealthy elites, promoted their policy interests while in office, while passing along a portion of the gain to a particularistic electorate from whom he "rented" his authority. (p. 1155)

Although it based itself on the voting support of the least privileged segment of the population of the industrial city, the machine played a basically conservative role in the political life of the city while fueling its economic growth. Machine politics encouraged urban growth through large-scale public infrastructure investments directed toward the major commercial and industrial districts. At the same time, the machine managed to control ethnic working-class neighborhoods through its strategy of small favors to needy supporters, mixed with more substantial rewards for local business and political leaders who acted as machine boosters in those neighborhoods.

Overall, by its lack of class orientation, its emphasis on concrete material rewards, and its support for traditional neighborhood ties, the machine form of political organization provided an important benefit to urban business interests by blunting or diffusing potential class conflicts between urban workers and business elites.[3] Thus, rather than acting as a working-class political party based on differences in class interest, the machine served an important integrative function in big industrializing cities that were ripe with the possibility of serious class conflict. It did this by playing the dual role of buffer and broker in urban politics. At the input side to city politics, the machine provided an organized channel of access to municipal government; at this point, it absorbed and buffered the energies of the newcomers to urban politics with its institutionalized system of exchange of public favors for private contributions of money, time, and energy. At the output side, it acted as the broker and distributor of governmental rewards, directing the largest public investments to the highest bidders, who generally were located in the business sector, and many smaller rewards to its supporters in the residential communities (Katznelson, 1976). This type of distributive politics was consistent with the exchange principle that underlay the functioning of the machine.

At the same time that it acted as a channel for socialization and for controlled participation by newcomers to the city, the machine also invested in those promising areas of growth in the city's economy and in its residential communities that were in turn willing to reciprocate and to invest in it by their contributions. This meant that the machine functioned as a decision center for public investment in the developing industrial city. The areas in which it invested were typically areas of spontaneous, unplanned development. Machine investments were made proportionate to the willingness and ability of such private growth centers to enter into exchange relations with it. In general, the machine fed off the growth of the industrializing city, and in turn it stimulated that growth (Glaab & Brown, 1967). Without rapid growth, the machine's consumption of monumental amounts of graft would have become a serious burden on the city's economy; indeed, the breakdown of the machine system during the Depression of the 1930s testifies to this.

The classic machine was both an organized citywide political network[4] and decision center in an expanding pluralist political system. Its extensive organizational linkages gave its leadership the opportunity to oversee the city as a whole and attain a perspective that was unavailable to any single interest group. This could have been of considerable value to the rational economic development of the city had the machine been oriented toward systematic planning of urban development. Yet its policy regarding public investment was essentially a matter of ad hoc response to spontaneous local development through the medium of exchange relations. Typically, the machine did not engage in long-term planning of urban physical or economic growth; rather its allocations of public resources reinforced growth tendencies already in existence or facilitated their further expansion. In short, the machine acted as an ad hoc decision center whose basically distributive policy provided reinforcement to the development of growth areas that engaged in its exchange processes.

The medium for this policy was a web of political-economic exchange relations through which selected organizational networks of the community connected with the organizational network of the machine. All of this took place on the plane of the community; no higher levels of government or business were involved. In sum, the political machine provided a central coupling mechanism between public and private interorganizational network processes, thus facilitating the flow of economic and political resources essential to local development, all at the horizontal level of the urban community (Laumann, Galaskiewicz, & Marsden, 1978). Vertical ties were to come later, with the development of the federal urban policy system and the complex of intergovernmental relations associated with the provision of programs and resources by this system to the local community (Milward, 1982).

URBAN REFORM AND THE
REORGANIZATION OF LOCAL GOVERNMENT

By the end of the 19th century a movement for reform of urban political life had begun to take shape, aimed at removing the machines from their positions of political dominance and replacing them with a more predictable and efficient form of municipal government. Seen from the perspective of its supporters, reform was a morally inspired revolt against political corruption undertaken by a broad range of respectable urban dwellers who sought to replace the machine's unscrupulous practices with accountability and efficiency in urban government. One familiar account of reform subsequently developed by social scientists pursues this same theme, picturing urban reform in terms of a confrontation between two contrasting value systems. Social historian Hofstadter (1955), a major source for this view, interpreted the drive for reform as the result of an inherent conflict between sharply opposed systems of political ethics: "One, founded upon the indigenous Yankee-Protestant political traditions, and upon middle class life, assumed and demanded the constant, disinterested activity of the citizen in public affairs . . . and expressed a common feeling that government should be in good part an effort to moralize the lives of individuals" (p. 9). Hofstadter sharply contrasted this value system with that of the European immigrants, in which family needs and personal loyalties ranked "above allegiance to abstract codes of law or morals" (p. 9).

Should any doubt remain about which side virtue resided on, political scientists Banfield and Wilson (1963), commenting on Hofstadter's portrayal, observed that:

> The Anglo-Saxon Protestant middle-class style of politics, with its emphasis upon the obligation of the individual to participate in public affairs and to seek the good of the community "as a whole" (which implies, among other things, the necessity of honesty, impartiality, and efficiency) was fundamentally incompatible with the immigrants' style of politics, which took no account of the community. (p. 41)

This paints a rather flattering portrait of the political ethics of the reformers, associating them with concern for the good of the whole community as well as such values as honesty and efficiency. The suggestion is that although their ethnic working-class opponents were parochial and self-seeking, the middle-class reformers were essentially above politics or any thought of personal gain, motivated only by the desire to replace the political machines with "good government."

An alternative view is presented by political historian Hays (1984), who argued that the middle-class-oriented public service ideology developed by the leaders of the mainstream of the urban reform movement did not represent accurately their actual class positions or their underlying political objectives. Hays made it clear that the reformers spoke in the name of the public interest and of democratic government primarily in order to attract the support of the urban middle class, which had often felt itself excluded from effective political participation under machine government. He demonstrated that rather than stemming from a broad middle-class movement, urban reform was typically initiated and led by members of the urban upper class.

Hays's (1984) analysis of the membership of the principal reform organizations and of the leadership of reform-oriented municipal research bureaus in such major cities as New York, Chicago, Philadelphia, and Pittsburgh indicates that those in the forefront of the reform movement were drawn from the ranks of prominent businesspeople and from the leading professionals of the time, representing the most advanced elements of their respective fields. The businesspeople came from "the most powerful banking and industrial organizations of the city . . . not the old business community" but the growth industries of their time, such as railroads, steel, coal, plate glass, and the emerging electric industry, "which had come to dominate the city's economic life." Their interests were associated with the introduction and expansion of modern industrial technologies, in short, with modernization and efficiency in an industrializing economy. The professionals too were modernizers in their own context, drawn from "the vanguard of professional life, actively seeking to apply expertise more widely to public affairs" (p. 59).

The upper-class nature of the movement for structural reform can be understood more fully by recognizing that although it was dominant, it was not the only approach to reform. Other efforts to reform city life, constituting a movement for social reform, placed major emphasis on improving the social and economic conditions of the urban working and lower classes. The leaders of the settlement house movement sought to provide material and educational assistance to the working-class and immigrant populations and "to influence government activities in a progressive direction" (Harrigan, 1985, p. 93) such as building public housing, providing unemployment insurance, combating child labor, regulating factory safety conditions, and the like.

Another approach pursued by a variety of social reformers ranging from settlement house advocates to socialists was to run for public office in order to directly shape public policies. Those who were elected often adopted policies of providing public services at reduced rates, in some cases putting the city in the electric utility or waterworks or streetcar business in order to make these services affordable to lower income citizens. Around the turn of the century, social reform mayors in a number of industrial cities also initiated or expanded free public park and recreational facilities and funded public school construction and programs of public relief for the unemployed. Overall, however, the major stream of reform was that associated with the structural reformers, whose influential positions brought superior funding and ready access to political officials and the press, in pursuit of reforms emphasizing changes in government institutions rather than in lower class social conditions.

Urban Growth and the Emergence of Structural Reform

Underlying the early 20th-century movement for urban political reform were a number of issues associated with the economic and social consequences of rapid urban growth. From the mid-1800s to the early 1900s, the population and industrial plant of American cities grew at an unprecedented pace, requiring a wide variety of new public facilities and services including paved streets, streetcar systems, bridges, waterworks, sewage systems, and electric utility plants. This provided ample new opportunities for enterprising businesspeople to invest in profitable municipal ventures. However, access to these lucrative opportunities increasingly lay in the hands of machine politicians who were able to arrange for the awarding of municipal contracts to construct these facilities, as well as franchises for their operation. By the turn of the century, over $2 billion had been invested in street railway systems, with another half-billion in investments divided between electric lighting plants and gas companies. In exchange, "contractors and developers were expected to return part of their profits to the political machines" (Shank & Conant, 1975, p. 25), whose operators in turn used this graft to keep the machine running smoothly, as well as for personal gain.

Graft was a major medium for oiling the party machine and reducing possible friction points that might stand in the way of political or economic enterprise in the early industrial city. Financially rewarding contracts for public works were awarded regularly to businesspeople in return for kickbacks; retention of the contract depended largely on maintaining payments to the machine (Savitch, 1979). Although businesspeople found it possible to accommodate to a moderate level of graft in exchange for these rewards, unbridled corruption threatened to reduce this system of finely balanced political-economic exchange to one of outright plunder, draining the profits of cooperating businesses. Machine excesses finally reached such dimensions that they came to be perceived as a serious economic threat by urban business interests. In particular, large corporate businesses, seeking to minimize their local costs as they pursued growth beyond their established urban base, began to pull away from contributing dutifully to the machine. Machine-sponsored practices such as writing of purposely vague regulatory ordinances, unpredictable selective enforcement by bribe-seeking officials, and arbitrary changes in the size of required payments led to intense dissatisfaction with the machine as urban-based business entered the phase of large-scale growth.

Because the continued growth of the city could be identified reasonably with protecting the profitability of business enterprise and hence its capacity to provide jobs and contribute to the municipal tax base, it was not a long step to identifying the general civic interest with the interests of urban business. Soon blue-ribbon committees of leading citizens, voters' leagues, citizens' unions, and research bureaus began to spring up, financed by local elites in banking, commerce, and industry. Together these organizations laid down the roots of interorganizational networks that ultimately extended nationwide, collecting information on municipal affairs, devising model city charters and political strategies for getting them enacted, and generally disseminating propaganda favorable to the reform cause.

The earliest national-level reform organization, the National Municipal League, was established by the First Annual Conference for Good City Government in 1894. By 1916 the league had developed a model city charter that included recommendations for council-manager government and city planning as basic institutions for reformed cities (Banfield & Wilson, 1963). During these two decades a network of reform organizations grew at both the national and local levels, effectively serving as a nationwide organizational infrastructure that would prove vital in sharing information, forming coalitions, and developing political strategies for the reform movement.

Among the first local reform organizations was the Bureau of Municipal Research of New York City, founded in 1906 and substantially financed with contributions by Andrew Carnegie and John D. Rockefeller. Within the next decade similar bureaus devoted to researching public finance, exposing waste, and proposing reforms that would promote public efficiency in municipal government were established in a growing number of cities. By World War I every major city and many smaller ones were linked into an urban reform network whose leadership was distinctly upper class and that drew on national business associations such as the Chamber of Commerce for material and political support (Hays, 1984).

The well-organized and financed activities of the reform movement succeeded in gradually forging a political alliance between urban business and professional leaders and the growing urban middle class. The new middle class were suspicious and fearful of the rapidly expanding immigrant population in the major industrial cities of the East Coast and Midwest. They associated the growth of this population with the growth in crime rates, alcoholism, poverty, violence, and political corruption that appeared to many to be aspects of the general decay of the quality of urban existence around the turn of the century. There seemed to be an urgent need to raise the standards of city life, although only in part to defend against the impact of alien populations. At the same time, with the continuing growth of an urban-centered industrial economy, local neighborhoods were threatened increasingly with encroachment by hazardous, polluting, and otherwise environmentally incompatible land uses, often due to the expansion of industry and of transportation networks impinging on residential areas.

Strong, reliable, clean government capable of rational regulation in various areas of concern—ranging from the social control of immigrant populations to the regulated zoning of different land uses—came to be perceived as essential to coping with problems of rapid urban development. The notion of scientific administration of industrial operations, associated with early 20th-century industrial engineering and the school of "scientific management" (Lewis, 1973, p. 74), was quickly snapped up among urban business and professional leaders and educated members of the middle class. They believed that the affairs of government too could be managed more efficiently through the application of principles of scientific administration to municipal organization, budgeting, planning, and personnel management; with this the discipline of scientific public administration effectively was launched.

Objectives of Structural Reform

The *structural reform* program can be understood as operating at two levels. On the level of ideological or symbolic politics, the structural reformers projected a positive self-image by associating themselves with the symbols of "public regardingness" and "good government," which they claimed were based in a political culture of fairness and efficiency. As political Progressives, structural reformers called for making city and state political systems more democratic through electoral devices such as the direct primary. They also pressed for the adoption of mechanisms for direct citizen participation in the legislative process such as public initiative, referendum, and recall.

In a number of respects these measures legally would open the legislative system to increased access by the members of local groups and communities and thereby contribute to its decentralization. However, as Hays (1984) observed, these innovations were intended more as symbols of reform than as practical means for democratization of the public policy process. They were useful to the upper-class leadership of the reform movement in fostering the popular appeal of reform and broadening its base of political support, particularly in the middle classes. The practical impact of these participatory innovations, although promising direct citizen participation in government, could be set aside easily by including provisions that required a high percentage—as much as 25% to 30%—of the electorate to sign the necessary petitions to set the process in motion (p. 69). Hays noted that Woodruff (1911) of the National Municipal League, the national coordinating body for the reform movement, stressed the symbolic nature of these participatory devices, observing that "their value lies in their existence rather than in their use" (p. 314).

At the level of practical politics, structural reform was not aimed at creating devices for direct democratic participation. Its primary objective was to make several basic changes in the organization of local political systems, which together actually would provide for greater centralization of decision-making authority in urban government institutions. These reforms were to be brought about by a number of organizational innovations, combined at appropriate points with new procedures of public administration.

Foremost among these was the centralization of municipal administrative authority, requiring consolidation in the hands of the chief executive of powers previously shared (as "weak mayor") with the city council. Second was the centralization of the system of political representation, involving a shift from local ward-based elections to citywide (at-large) election of city council members.

At minimum, administrative centralization would involve a move to "strong mayor" government, in which the mayor held significant powers of appointment and control over the heads of city agencies. At most, it would mean the replacement of mayoral government by council-manager government, in which a professionally trained administrator would act as the full-time managerial agent of the city council. In this case, there would be clear formal separation of administration from politics, a major objective of the reformers. The necessary powers for effective administration would be concentrated in the city manager; the political commitments and conflicts that are associated with the making of city policies would remain the business of the council. Closely associated with this conception was the assumption that council members, being elected at large, would be able to transcend narrow local interests and adopt a statesmanlike posture in their deliberations, thereby significantly reducing the level of politics in the making of municipal policy. Thus what the reformers were aiming at in their managerial model of local government was both the separation of politics from public administration and the general depoliticization of municipal policy making.

Along these lines, a related reform objective was the replacement of the system of partisan elections by one in which candidates were identified on the ballot by name only, rather than by party affiliation. Through the establishment of nonpartisan elections and through the holding of local elections at a time separate from national elections, the reformers hoped to break the grip of party machines on the local electorate and to encourage voters to consider criteria of technical competence and qualifications instead of political loyalties or party labels. One other major objective was the elimination of the practice of distributing positions in government agencies on the basis of patronage as a payment by the victorious party to its loyal supporters. Here too the reformers raised the banner of performance qualifications and technical competence as a basis for government service, seeking the establishment of a personnel system based on tested abilities—a merit-based system of civil service, rather than one based on political loyalty and favoritism.

Impacts of Structural Reform

Ideally, what the structural reformers wanted was a professionally managed fiscally efficient government organized along the lines of the model of a modern corporate enterprise, with the council operating as a board of directors and the city manager as the council's professional administrative agent overseeing and coordinating the work of the city's agencies in the delivery of public services. Overall, the objective was to break the machine's electoral power and to foster structural modernization of city government through the reorganization of political institutions. The benefits would include the cutting of expenditures and hence of taxes for public services and the reduction of political conflict between social classes. In particular, if the low-income tenement wards' vote could be minimized through a combination of nonpartisan elections and at-large voting, the socialists and other social reformers would have less chance of successfully demanding expensive public programs to improve conditions of the urban working class.

The specifics of what the reformers achieved varied mainly with the size and age of the city; in smaller and newer, often suburban municipalities in the

North, and throughout most of the newly urbanizing South and West, they had a rapid series of successes, "resulting in a host of small-city commission and city-manager innovations" (Hays, 1984, p. 66). In the large industrial cities with significant concentrations of immigrant and working-class populations, the best they usually could achieve toward formal reorganization was the strengthening of the powers of the mayor. At the same time, they continued to work through the municipal research bureaus and civic leagues that had been established in these cities and that provided a certain degree of reform influence on local government policies and practices in such areas as civil service and budgeting.

Their other goals of restructuring urban government to a formal managerial system and introducing citywide nonpartisan elections at first tended to escape them in the larger cities where the machine was generally strongest. Lower class voters concentrated in inner-city electoral wards often identified with the political style of machine functionaries who provided personal favors and ethnic recognition. They also supported the decentralized, relatively informal delivery of public services on a ward basis. The popularity of this neighborhood-level system of administering services contributed significantly to the difficulty of achieving structural reform in big cities. Nevertheless, reform ultimately had an impact on various aspects of governmental practices in most cities, ranging from merit-based personnel recruitment to the enlargement of executive powers. In this it was helped considerably by the experience of the Great Depression, the severe crisis of the national economy that lasted from 1929 through the late 1930s.

By the middle of the 1930s local economies around the country had collapsed, and there were sharp reductions in the revenues of municipal treasuries resulting in fiscal default in over 1,400 cities during the decade (Alcaly & Mermelstein, 1976, p. 213). This in turn meant the loss of the public resources that the machines had used so long to establish and maintain their web of exchange relations with urban business and residential communities. In particular, the machines no longer could distribute their informal type of welfare to the working-class families and communities who were their major voting constituents. In their place, the New Deal administration of President Franklin Roosevelt established public assistance programs that effectively nationalized the former welfare function of the machines (Harrigan, 1985). Regardless of their earlier popularity, the machines found themselves unable to compete with federally funded, professionally administered social welfare organizations whose programs were available to eligible recipients as a matter of legal right, rather than political exchange or patronage.

By the 1950s, viewed simply in terms of formal organizational change, the machines generally had lost their former position of political-administrative dominance, and reform had scored considerable success in the area of local government administration. In larger cities that were unwilling to embrace fully the reform ideal of a council-manager system, reform efforts went forward toward strengthening the managerial powers of the office of mayor. In general, the trend has been to provide the mayor in such cities with a chief administrative officer or at least a special administrative assistant to provide executive liaison and oversight for the city's agencies. In addition, there has been widespread adoption of civil service and a new emphasis on professional qualifications for leadership and staff positions in municipal agencies, so that most American cities of any importance today bear the imprint of the reform impulse toward modernization of public bureaucracy (Boesel, 1974).

URBAN GOVERNMENT AND POLICY
IN THE BUREAUCRATIC CITY-STATE

Although urban reform has had considerable success in changing the formal organizational structure of local governmental systems, full understanding of the impact of reform cannot be gained simply by analyzing changes in formal structure. Further examination indicates that the restructuring of urban government has led actually to the transformation of machine governance, rather than to its elimination. In place of the old politically based machines, city government now revolves around public-service-providing agencies that function in various respects as bureaucratically organized machines (Lowi, 1967). To understand this better, it is useful to examine briefly some of the basic similarities and differences between old and new machines.

Probably the major distinction between the traditional party machines and the new bureaucratic machines is that the new machines are functionally based, with the primary source of their power being the organization of specialized capacity for some particular public service function, such as education, housing, or urban planning. By contrast, the power of the old machines was territorially based, grounded in the intensive organization of political support in ethnic and working-class neighborhoods. In addition, the new machines have developed significant connections with agencies performing similar functions at the federal and state levels, reflecting the growing influence of vertical intergovernmental relations.

With regard to similarities, the new machines have taken the place of the old in the provision of traditional services, such as police and fire protection, street maintenance, sanitation, and water supply. In addition, they have added new services such as urban planning and renewal that were beyond the operational routine of the traditional machines. Although they are headed by career bureaucrats rather than political cronies, these service-delivering organizations are, like the old machines, relatively self-governing structures of power that are able to shape public policy with little accountability to either higher authorities or the general public. Their bureaucratic administrators may be more efficient and technically rational than were the old machine bosses, yet they too are political in their own way. Although the public agencies that they head operate on the basis of formal authority rather than the mobilization of voters' support, these "new machines" resemble the old in that they too are "relatively irresponsible structures of political power" (Lowi, 1967, p. 87).

This is probably an outcome the structural reformers did not themselves foresee. A thorough response to the question of what happened to urban reform-through-reorganization therefore must go beyond the usual focus on formal restructuring of municipal government to the substantive issue of the actual control of public organizations. In the era of bureaucratized city government, each municipal agency actively delineates and seeks to protect or possibly expand its area of functional scope, or *organizational domain,* as against other agencies (Warren, Rose, & Bergunder, 1974). This is an important aspect of the new bureaucratic politics of local government. As a result, modern city government has become fragmented into an array of professionalized public bureaucracies that stand as separate islands of functional power, often operating beyond the control of the mayor, the council, or the general public. Consequently, in any given city, there are now many more new machines than there once were old machines.

After more than a century of weak central government in American cities, structural reform introduced formal centralization through organizational changes in electoral and governmental institutions. Yet, in the long run, reform tended to encourage the functional decentralization of local government. It did so by facilitating the creation of relatively autonomous bureaucratic public organizations that are resistant to central control while eroding the old machines that in their own day had provided informal functional centralization of city governance. In short, although its purpose was to strengthen the central administrative authority of the city's chief executive, reform instead resulted in the elevation of the municipal bureaucracies to positions of nearly independent power. Political scientist Lowi (1967) summed up the results of this process in his observation that "the legacy of Reform is the bureaucratic city-state" (p. 86). As he remarked:

> Politics under Reform are not abolished. Only their form is altered. . . . Destruction of the party foundation of the mayorality cleaned up many cities but also destroyed the basis for sustained central, popularly-based action. This capacity, with all its faults, was replaced by the power of professionalized agencies. But this has meant creation of new bases of power. (p. 86)

Lowi performs an important service by raising the issue of the actual substantive consequences of reform, making the case that the old centralized political machines have in fact been replaced by poorly controlled decentralized bureaucratic machines engaged in their own brand of agency politics. His argument points up the conclusion that machine politics has not been eliminated; it has only changed form. If such is the case, what precisely is the new form of machine politics in the bureaucratic city-state?

The Structure of "New Machine" Politics

As I noted earlier, the traditional machine was an intermediary organization that connected multiple interests from the business community and the ethnic neighborhoods to the public service delivery agencies of the city government. Generally the machine took an active role in organizing politically inexperienced newcomer groups, rather than waiting for them to do so independently. With the decline of political machines, a variety of agency-clientele relations have sprung up between public bureaucracies and the client groups that they serve. Frequently these relationships are unmediated, occurring directly between agencies and clientele groups—ranging from business associations to neighborhood organizations—that today are more often self-organized (Allensworth, 1975) and actively seeking agency access. Other agency-clientele relationships tend to be mediated by lawyers, local area politicians, advocate planners, or others who act as knowledgeable representatives of the clients, particularly those clients who lack strong independent organization, such as the poor.

The proliferation of public agency-client relations is not surprising because "bureaucracies are formed and exist to represent, protect, and advance particular interests" (Allensworth, 1975, p. 29). One of the major sources of the power of public bureaucracies is their ability to foster such relations with their clients and to mobilize support from clients and sympathetic members of the general public who have special interest in the agency's activities and services. In serving the interests of their clients, public bureaucracies indirectly serve their own interests; they find support for the claims they make on the government's budget for purposes of expanding and improving their programs by adding staff, new technologies, and the like. This should be kept in mind as a basic aspect of the new politics of bureaucratic government.

Perhaps somewhat less expected is the degree to which the personnel of the public agency—the career administrators, professional staff, and unionized employees—have themselves emerged as an independent set of interest groups concerned with the design, staffing, and delivery of agency services, zealously guarding from external challenge their established service routines and the budgets that pay for them in a manner that can be described only as political. This is yet another aspect of the bureaucratic politics of the new machines.

In this perspective the bureaucratic city-state can be understood as a governmental system characterized by the growth of organized interests within the public agencies themselves, parallel to the organization of outside clientele interests. Within the bureaucracies the job protections of civil service and the development of powerful associations of public employees—police, firefighters, sanitation workers, teachers, and the like—have tended to produce well-organized service provider groups with strongly vested interests in the control of agency programs and service delivery routines. At the same time, a wide variety of groups outside the public agencies—including business organizations, professional associations, church groups, racial and ethnic organizations, tenant organizations, homeowner organizations, and others—have organized themselves in order to gain access to these agencies and to related governmental institutions. Being self-organized, they are considerably more free to set their own agendas in dealing with the new machines than were the ethnic immigrant communities, which were actively organized by the old machines primarily in order to control their votes. The contemporary tendency to self-organization and self-assertion of group interests has been encouraged by the changed style of politics associated with the new machines. As Piven (1976) observed in examining some of the political consequences of the bureaucratization of urban government:

> Today public goods are distributed through the service bureaucracies. With that change, the process of dispensing public goods has become more formalized, the struggles between groups more public. . . . [W]hile we may refer to the schools

or the sanitation department as if they are politically neutral, these agencies yield up a whole variety of benefits, and it is by distributing, redistributing, and adapting these payoffs of the city agencies that urban political leaders manage to keep peace and build allegiances among the diverse groups in the city. (p. 134)

In such circumstances, no group that understands the nature of the new bureaucratic politics will hang back waiting to be invited to assert its interests. Only groups whose members are lacking in political resources or experience or who are impeded by social or political bias will be slow to organize and enter the circle of claimants for government services or largesse. To delay in entering the contest means leaving the channels of access to be occupied by others and possibly being kept from fully sharing in agency benefits by other groups' earlier development of organizational relations and of official recognition in the form of budgeted agency services.

For the public agency the proliferation of assertive interest groups creates a dilemma. Every agency seeks to cultivate a clientele of recipients of its services that can function as an active constituency ready to lend it political support in speaking positively of its programs to public officials and a wider public and in backing its budgetary requests. Typically, well-organized clients are prepared to act directly as a supporting constituency for the agency; others with less political experience often have relied on sympathetic advocates or interested professionals to represent their viewpoint. Yet, after a certain point, the proliferation of clients begins to challenge the agency's capacity to deliver services adequate to the size and variety of demand, outweighing whatever value the clients may have as active constituents (Gilbert, 1970).

Ultimately, the problem of demand overload on individual agencies translates into a problem of overload on the city's fiscal resources. Individual agencies may deal with demand overload by "skimming" their clientele (catering to the less disadvantaged and the more conventional who are easiest to serve and who often have more political clout) as well as engaging in other strategies of controlling or delaying demand. But as potential client groups become better organized, evasion of their demands becomes more difficult. The problem becomes particularly acute in a time of simultaneous political and administrative fragmentation, such as that during the War on Poverty in the late 1960s and early 1970s. In such a period, the moderately pluralistic governmental system that long has been characteristic of larger American cities enters into a phase of multiplying group participation and conflict sometimes described as *hyperpluralism* and tends to become fiscally overloaded (Wirt, 1973).

URBAN POLICY AND DEVELOPMENT
IN INTERORGANIZATIONAL/POLICY PERSPECTIVE

Compounding these problems in recent decades has been the rise of urban political coalitions with strong interests in the expansion of public services. Such coalitions confront the typical big-city mayor as the one governmental official ultimately held responsible for the policies and management of the city's agencies. According to Yates (1979), the first of these coalitions is a *service-demanding coalition* that includes neighborhood groups, urban bureaucrats and professionals, and municipal employee unions interrelated on the horizontal level of the community. Although there is often conflict among these groups, there is a commonality in their demands for increased or improved services. The second coalition that confronts city hall consists of *service providers,* tied together on the vertical dimension that connects functionally related service delivery agencies at the three levels of the American intergovernmental system. Like the first coalition, this one includes personnel of city agencies, particularly career bureaucrats and professionals. In this case, however, they are linked not to local interest groups but to state and federal bureaucrats whose administrative activities govern the local allocation of services. The mayor is thus faced with powerful political coalitions on both the demand side and the supply side who support the expansion of public services. Significantly, a central role in both the service-demanding and service-providing coalitions is played by public agency professionals who "seek to extend their domains in service delivery"; they are supported in this by the municipal unions "because more services means more jobs" (p. 64).

By the early 1970s the convergence of interests between the service-demanding and service-providing coalitions in American cities tended to act as a powerful force for increased spending for public services at the same time that it kept service delivery programs and procedures beyond the control of city hall. Yates (1979) observed that "on this account the mayor was faced with coalitions that blocked strong central control" (p. 65) by him in major areas of urban policy. These coalitions do not operate only on the immediate horizontal level of the urban community; the coalition of service providers, for example, involves organizational linkages between bureaucrats involved with programs in housing, education, or health care in functionally related agencies at local, state, and federal levels. In contrast to the era of the political machine when urban interest relations were confined essentially to the level of the local community, an extralocal vertical dimension reflecting the importance of urban intergovernmental relations is today a significant aspect of the administration of public policy in American cities.

This vertical dimension is involved not only in the administration of already established policy but also in the basic processes by which policies are developed, including policies that provide legal authority and funding support for programs of importance to urban areas. The programs for urban renewal and the War on Poverty, for example, were funded primarily at the level of the federal government rather than at the level of the local communities in which they operated. In the case of the antipoverty programs, they were developed initially by federal officials and service professionals who believed that it was necessary to depart from the established program routines of local service bureaucracies in order to better address the needs of the urban poor. In these respects, all of these were truly programs of the vertical dimension, with a legacy that can be traced to the New Deal of the 1930s.

In order to introduce an interorganizational/policy perspective clearly based on historical experience, this chapter has tracked the major changes in urban policy process from the political machine of the late 19th century to the modern era of the bureaucratically managed city. I have noted the shift of policy focus from working-class immigrants and upwardly striving business and professional interests toward middle-class residents and a new establishment of corporate and professional leaders in the reform era. But equally as important were the organizational changes resulting in a basic transition from the city governed through a broadly decentralized machine to the concentration of functional authority in centralized municipal bureaucracies.

Of particular interest, from the perspective of this book, is the fact that organizational changes in urban government and politics proved as significant as the changes in policy focus. These organizational changes affected both the participation of citizens in urban politics and the balance of influence between competing urban policy coalitions by shifting political control over the policy process toward the relatively higher class participants prominent in the reform coalition.

The point here is that the specific form in which the powers of government are structured makes a difference to both urban political and economic development. Such was clearly the case when urban politics and policy making could be analyzed essentially on the plane of the community (the horizontal level). A major theme of this book is that how the powers of government are organized still makes a difference. Since the New Deal, governmental power structured on the vertical axis—typically transmitted through intergovernmental programs involving interorganizational ties between related federal, state, and local agencies—has become ever more significant to urban and related social policy.[5]

The focus in the next part of the book will be on tracing several key developments in urban policy since the 1930s and their relationship to changes in the American federal system, viewed as a system of changing interorganizational relations. This examination is aimed at showing that the operation of the intergovernmental system has significant impacts; specifically, that it can serve to open up or close off important possibilities for urban social-economic and spatial development depending on how the system is organized and what policies it emphasizes.

Cooperative Federalism and Intergovernmental Relations

During the 19th century there was a long-standing practice of separation between federal and local government responsibilities that came to be known as *Dual Federalism*. With the crisis of the 1930s' Depression, historical precedent was set aside and massive economic resources flowed from the federal government directly to cities in need of assistance. At the time, it was assumed that the federal government's aid to cities was to be no more than a temporary measure; hence its intervention through intergovernmental programs such as public works and public housing was accepted. It is now clear that these federal-local program linkages were only an early stage of long-standing relationships that were to multiply in size and significance until the late 1970s. In the postwar era, suburbanization on the one hand and urban renewal and problems of poverty on the other produced increasing demands by municipal governments on Washington for assistance in dealing with their economic and social problems.

The net results of these developments were reflected in major changes in the American federal system. In place of an intergovernmental system characterized by clearly differentiated powers and responsibilities between national and state governments, a new structure of Cooperative Federalism came into being.

The new system has been characterized by changing relations of interdependence between federal, state, and local governments (Reagan & Sanzone, 1981). Across a wide range of policy areas, "domestic policy and programming today involve all three levels of government simultaneously," so that contemporary federalism entails "intense, regular contacts" among officials representing different governmental levels (Hahn & Levine, 1984, p. 30). In a shared power system of that type, no single governmental level holds all the power.

Much of the interaction between related agencies becomes a matter of bargaining, through which policy and program agreements are developed regarding the respective roles to be played and the resources to be exchanged between governmental levels in the process of policy implementation.

Consequently, although urban policy issues have become nationalized in that they have obtained recognition as problems of national scope, no uniform national policy solutions have been found, nor is the evolving system of intergovernmental relations by any means one that faithfully follows some policy line established in Washington. The periodically changing intergovernmental system remains based on negotiated, shared power. It is a system that during recent decades has become more centralized in some aspects yet that remains decentralized in many others, reflecting an ongoing contest for control over both the definition and implementation of public policy between national and subnational levels of government.[6]

As a structure for policy making, the contemporary intergovernmental system involves multiple networks of public agencies and coalitions of private organizations; the organizations in each network are interlinked based on their mutual involvement with particular programs and policy issues. In order to carry out a given policy, consent must be obtained from the various interests that compose an organized domain of policy interest. When policy conflicts arise between these different interests, they are settled typically "through processes of reciprocal exchange, negotiation, intermediation, and bargaining" (Rondinelli, 1975, p. 261). Consequently, the process of policy implementation involves the working out of agreements both among agencies at different levels of government and among related clientele groups and policy-interested political coalitions.

Several major policy coalitions since the 1930s will be identified as they relate to changes in the structure of the intergovernmental urban policy subsystem and to shifts in its policy emphasis: the New Deal coalition originating in the 1930s, the urban renewal progrowth coalition, which held sway from the mid-1950s to the mid-1970s, and the coalitions in favor of a New Federalism since the mid-1970s.

Much of government's involvement in the restructuring of urban America over the past half century can be understood by reference to the emergence of influential political coalitions with interests relevant to the urban development process. Frequently these coalitions, combining private and public interests, have been assisted by the activities of public policy entrepreneurs who created "new governmental bases for exercising new powers which none of these actors and interests could otherwise have exercised on its own"

(Mollenkopf, 1983, p. 4). Such policy entrepreneurs include activist mayors and urban-oriented members of Congress, as well as urban program administrators at both the local and federal levels. One aspect of the role of these entrepreneurs has been to sponsor the creation of public programs in response to perceived or expected demand from urban interest groups. Establishment of new programs then may lead to the organization of previously inactive groups that, as eligible program beneficiaries, will be likely to support further program development and expansion.

Thus, where redevelopment coalitions did not already exist, policy entrepreneurs systematically have encouraged their formation with the promise of programs that would serve their interests. In turn, this has led to the emergence of new policy-interested lobbying organizations and coalitions (Haider, 1974) and to pressures from those organizations for creating or expanding public agencies and programs that would implement policy suited to coalition interests.

In connection with these developments, urban interest groups have engaged in seeking coalition partners, not only at the local level (as historically had been their practice) but increasingly also at the national level. These new relationships can be approached in terms of the concepts of *coalition* and *federation* that are basic to the emerging study of interorganizational relations in contemporary society and its urban communities. To place these terms in context, it is useful to apply Warren's (1967) typology of interorganizational fields, which is a spectrum of four different field types at the community (or horizontal) level.

The least structured field type is the *social choice* model, in which relations between organizations in the specified field are essentially of a free market nature, with the organizational units clearly independent of each other in their decision making. (In historical perspective, this approximates the situation that typically preceded the development of stably institutionalized interorganizational relations in American cities prior to the late 1800s.) At the other end of the spectrum, there is the highly structured *unitary* type, in which field relations in a given policy area or subarea have been subsumed into a single hierarchically structured coordinative organization, such as a city health department. The interorganizational field here is collapsed into a bureaucratic entity whose components have little autonomy, acting simply as operational arms of the guiding central bureaucracy. Taken to its logical extreme in the urban setting, such would be the public agency model for a city governed by highly centralized bureaucratic organizations. Along these lines, some experimentation with unitary superagencies was in fact attempted in urban government during the 1970s; notably though, the attempts at top-down bureaucratic reform did not meet with much success.[7] For current purposes here, then, the most interesting and relevant interorganizational field types for consideration are the intermediate models: the *coalitional* type and the *federative* type, as defined by Warren (1967).

At the level of the local community, a coalitional context exists in the community's organizational field when there is a group of organizations (for example, local hospitals) that are willing to cooperate to attain a desired objective, such as exchanging needed resources—information, personnel, technology, clients—with one another or increasing their influence with public organizations that might provide needed resources or the funds to obtain them. Although each of the participating organizations has its own goals, as a coalition they collaborate around specific goals that they share. Decision making as to what steps must be taken to achieve these goals proceeds on a lateral basis among the cooperating organizations themselves with no authoritative unit at a higher level formulating policy in the name of the coalition. In effect, decisions are made through a process of bargaining and negotiation on the same level between the participating organizations in the coalition.

A basic federative context for decision making is present when the participating units establish an inclusive central unit to which they delegate authority to act on their behalf. Local examples of such an arrangement include a council of social agencies and a conference of churches. Although member units retain much of their autonomy, at least a "moderate degree of collectivity orientation—consideration of the well-being of the inclusive organization—is expected" (Warren, 1967, p. 405). Thus the development of federative relations among organizations involves the surrender of some degree of autonomy by member units to a central unit that they place above themselves in exchange for what resources and coordination it can provide to them. Although traditionally confined to the horizontal level of the community, local federative relations foreshadow an important aspect of the vertical relationships that today link local communities to agencies of state or federal government or to private sector organizations of the larger society (Scott & Meyer, 1991).[8]

In the context of the national society, significant federative arrangements are an integral part of the fundamental structure of American government. Under the traditional system of Dual Federalism this involved what originally seemed a fairly clear-cut and stable division of responsibilities between the federal government and the state governments. After 200 years of evolving adaptation to changing social and economic circumstances, the United States is today in an era of intergovernmental relations consisting of a variety of often uncertain and changing federal-state-local relationships, typically involving vertical federative linkages between agencies jointly involved in implementing designated federal programs (Milward, 1982).[9]

Over the past half century the precise nature of federative arrangements between these different governmental levels has become quite fluid, with accompanying changes in the connotations of federalism. To a considerable extent what has been involved is the issue of the relative powers of the federal government, charged with making inclusive policies for the welfare of the national collectivity, as against the powers and prerogatives of the states. Of course any important shifts in the balance of federal and state power hold important implications for the development of urban communities, which legally are creations of the states. At the same time, the rapid postwar evolution of collaborative relations between national and subnational governments contributed to breaking down the walls earlier erected under Dual Federalism between federal and state responsibilities. In particular, this has included innovative developments in direct federal-local program relations, although recently these have been curtailed by cutbacks in funding and changes in the structuring of intergovernmental grants, giving greater emphasis to the role of the states (DeGrove & Brumback, 1985; Wood & Klimkowsky, 1985).

In Part II of this book I review developments in urban policies in relation to changes in policy coalitions and policy objectives and to developments in the federative structure of the intergovernmental policy system that have produced a mixed and more flexible system that can be characterized as one of "articulated variety" (Kirlin, 1978, p. 85).[10] With regard to policy system structure, I will pay special attention to the operation of the major intergovernmental devices created for delivering urban program resources to central cities: *categorical grants-in-aid*, narrowly targeted to meeting a specific need or purpose; as well as two less restrictively targeted types of grant: *special revenue sharing*, more commonly referred to as *block grants*; and *general revenue sharing* (Reagan & Sanzone, 1981). In regard to this, I will discuss the performance of these program delivery mechanisms in implementing particular types of urban-relevant policy and examine how changes in policy emphasis have been related to changes in the structuring of program delivery.

Political Coalitions and Urban Policy

Historically, different policy coalitions have sought to organize the intergovernmental system in line with their own policy preferences. To the extent that its organization reflects their efforts, the system is not neutral but is structured to facilitate certain types of policy initiatives and to constrain or impede others. Since the 1930s several basic types of urban policy have been associated with the contemporary intergovernmental system. The two that are of major concern here are developmental policy and redistributive policy.

Developmental policies are essentially public strategies to improve the productivity and competitiveness of private business and industry, typically through improving public physical infrastructure as in urban renewal or the upgrading of transportation or utility systems.

Redistributive policies usually have been strategies to improve the condition of low-income and needy groups in the community (Peterson, Rabe, & Wong, 1986).[11] Redistributive policies tend to shift resources from middle- and upper-income taxpayers to these groups, particularly through welfare and social service programs targeted to the disadvantaged. Much of the discussion in the next part of the book is directed to the issue of the fitness or capacity of the established intergovernmental system and its several modes of delivering resources (categorical grants, block grants, and revenue sharing) for the implementation of redistributive versus developmental policies.

In regard to this issue, my thesis can be stated in two basic points. The first is that there is a reciprocal relationship between the intergovernmental system and policy-interested groups and coalitions in its environment. Changes in the structure of the intergovernmental system are influenced by changes in the composition and objectives of the leading coalitions of organized policy interests. In turn, those policy interests are themselves influenced by the activities and interests of governmental organizations within the intergovernmental system, in a sequence typically as follows.

In general, newly dominant political coalitions introduce policy thrusts that require the development of organizational capacity in government to administer the new policies. This leads to the creation of new public agencies or programs involving distinct but not fully defined administrative missions and interests. Over time, specific agency functions are negotiated within the context of working relations with other participating agencies at various levels of relevant policy networks in the intergovernmental system. The result is a public policy system that during the past half century has been the setting for significant changes in interorganizational relations. Its participating agencies have established strong interagency linkages between the policy system's federal and urban levels. At the same time, these agencies have reached out actively to encourage the organization of supportive interest groups in the private sector and have then mobilized the support of such groups for expansion or enhancement of agency missions and budgetary requests. A complicating factor is that all of this occurs against the bureaucratic resistance and inertia of already established agencies, which embody the policy thrusts of earlier dominant interest coalitions. Thus the existing policy system at any given point in time is really a mixed structure that combines new forms and earlier ones, rather than being a fully consistent translation of current policy intentions into administrative practices and interrelations.

This leads to my second point. Built into the American intergovernmental system at a very deep level of its structuring and supported by both traditional administrative relationships and contemporary political interests is a long-standing conservative bias against certain types of policy—most distinctly against redistributive policy. The basic organization of the federal system leans in the direction of decentralization, involving multiple checkpoints for decision and implementation and allowing for discretion at the subnational level that can block or distort policy intentions established at the national level (Pressman & Wildavsky, 1979). This structural bias is most likely to impede those policies whose major support comes from relatively less influential coalitions in social and economic terms, as is generally the situation for minority and low-income group interests. Such was the case, for example, for the redistributive programs of the War on Poverty, which often operated against significant resistance from established local political party and public agency interests and which effectively were dismantled after less than a single decade of operation. In contrast, developmental policies such as urban renewal have in one form or another been permitted to run their course over the last several decades, contributing significantly to the physical and economic restructuring of many older central cities while providing special benefits to major business and institutional interests in those cities.

Is it possible to reshape urban policy so that a reasonable balance of policy emphases is produced with regard to continuing needs both for development and for redistribution? Throughout this book this question will be examined in relation to ongoing changes in the intergovernmental policy system. A basic model for federally supported urban renewal in the postwar period has been one involving partnerships between large-scale public and private organizations[12] for economic revitalization focused on central business districts. At one level—for example, the development and application of new technologies (e.g., telecommunications and biotechnology) as postindustrial engines of urban economic growth[13]—some variation of the public-private model still may be applicable, now emphasizing local, rather than federal, partnerships between municipal governments and entrepreneurial investors. At a more modest level, however, is there some significant role in further urban development that can be played by organizations operating at a smaller scale, with not so high-tech an orientation, emphasizing the development and application of underutilized local manpower and community organizational resources?

If, at least for the near future, limited public resources are assumed, is there perhaps a developmental niche for community organizations operating as nonprofit structures of the relatively untapped voluntary sector? What role can community groups play in response to persisting problems of local underdevelopment and poverty that have eluded previous redevelopment efforts conducted at the level of large public and corporate bureaucracies? As against the elite public-private growth coalitions that have dominated the urban renewal scene, what are the possibilities for working coalitions between community groups, local public agencies, and pragmatic urban political leaders in a number of areas of mutual interest, such as housing rehabilitation and small business development in neighborhoods? Finally, in view of the fiscal impact of federal program cutbacks at the local government level, what has been the response of the states, and what are the types of public-private partnership that have been emerging locally on behalf of urban development?

NOTES

1. On interorganizational networks, see also Cook (1977); Knoke (1990); and Laumann and Knoke (1987). For seminal concepts underlying interorganizational analysis, see Levine and White (1961).

2. Examples of intergovernmental cities during the past two decades include Baltimore, Boston, Buffalo, Chicago, Cleveland, Detroit, Newark, New York, Philadelphia, and St. Louis (Burchell et al., 1984).

3. See Katznelson (1981), regarding the significant role played by urban machines in separating the "politics of work" and related broad working-class concerns from a narrowly focused "politics of community" that was systematically promoted by the machines (p. 193).

4. In the case of New York's Tammany Hall, for example, an encompassing organizational framework was provided by the machine's "network of district political clubs" that extended "throughout the city," joining otherwise disparate ethnic neighborhoods in complex political coalitions (Shefter, 1976, p. 35).

5. Regarding intergovernmental relations and their connections with urban policy and politics, see Brown, Fossett, and Palmer (1984); Burchell et al. (1984); Haider (1974); Mollenkopf (1983); Peterson, Rabe, and Wong (1986); and Warren (1985).

6. See Anton (1984). This is a critique of what Anton refers to as recent "stereotyped images" of federal centralization and dominance over domestic policy. His extensive review of the intergovernmental literature indicates that both state and municipal governments generally have been "able to bend federal program requirements to their own ends," in part because federal policy intentions are "typically flexible enough to accommodate a wide variety of local uses, providing plentiful opportunities to mix federal and local dollars" (p. 44).

See also Kettl (1988), regarding issues of decentralization and of the "conflict between federal control and state-local discretion" (p. 65) over the administration of federal grant programs. See Robertson and Judd (1989) on the consequences of the decentralized grants strategy first adopted during the New Deal, leading to problems of fragmentation and incoherence in policy implementation since then. For a useful theoretical perspective on interorganizational problems associated with policy implementation in federal systems, see Berman (1978). Berman observed that the execution of federal policy involves certain problems of "macroimplementation" because it is carried out in a "loosely coupled" setting composed of multiple relatively autonomous organizational actors, and it "passes through and is transmuted by successive levels of implementing operations." As a result, "the effective power to determine a policy's outcome rests with local deliverers, not with federal administrators" (p. 157).

7. See Hawley and Rogers (1974), who noted in their introductory chapter that in New York City, for example, Mayor John Lindsay in the early 1970s "consolidated the very fragmented fifty-odd agency machinery of city government into eleven superagencies" (p. 18). Lindsay's objective was to overcome the fragmentation of New York's municipal bureaucracy through closer planning and coordination among traditional agencies with functional capacities related to major urban problems or needs such as poverty, housing, economic development, and so forth. In each functional area the superagency model was intended to bring together the relevant existing agencies as departments of a superordinate agency, organized as a formally structured hierarchy. However, as Hawley and Rogers reported, many of the superagencies were not truly unitary but "merely collections of particular operating agencies that just happened to be under the same administrative umbrella," and did not foster an improved process of program implementation. Moreover, as formal hierarchies, they "often proliferated more bureaucratic layers between the citizenry and top city officials than had existed before, making for less rather than more political accountability" (pp. 18-19).

The response of Hawley and Rogers (1974) to this issue is of interest. They placed the superagency approach within the category of top-down strategies for improved integration of the service delivery systems of urban government and compared it with proposals for bottom-up change calling for increased citizen control of municipal bureaucracy. Their own approach is to endorse neither strategy to the exclusion of the other, but to seek an "appropriate mix of change strategies" (p. 21) that would employ aspects of each and also allow for use of alternative means of providing governmental services to encourage innovation and improved service delivery. These observations, made two decades ago, retain their relevance and still deserve consideration with regard to service delivery at the local level of government.

8. Scott and Meyer's (1991) review of Warren's (1967) work on interorganizational field types directs our attention to its relatively overlooked applicability beyond horizontal (single-level) interorganizational relations, such as those confined to the level of the local community. My own discussion follows their suggestion, applying Warren's federative conception to the analysis of vertical (hierarchic multiple-level) relations between community social units and higher level extracommunity organizational systems, for example, at state and national levels.

9. Regarding the importance of program as an organizing element in the intergovernmental system, Milward (1982) observed: "The granting of funds for a program in health, welfare, transportation, energy, or housing acts as a great chain, holding an enormous number of individual actors and organizations together—from the granting authority in Washington to the 50 states to thousands of localities where the services are delivered" (pp. 460-461).

The complexities of these policy-based program relationships indicate that "not one or even two or three organizations [are] involved in their administration but literally thousands" (Milward, 1982, p. 461), and not just public agencies but also private or nonprofit organizations are brought into the network through contract relationships to help in the provision of designated program services. In this perspective, it is clear that individual organizations are insufficient as the unit of analysis of policy implementation. More appropriate, as Milward argued, is the concept of the *interorganizational policy network,* which is composed of functional linkages between related agencies concerned with the intergovernmental delivery of a particular program or package of programs. It is around such networks that diverse private organizations and individuals tend to gather owing to their common interest in a given program.

10. Kirlin (1978) suggested that since the 1950s the major new strategic option emergent in the complex intergovernmental system has been what he called "articulated variety"—which he contrasted to several more familiar options: full autonomy of governmental units; classic Dual Federalism ("division of responsibilities"); and centralized or unitary federalism. For him, the advantage of articulated variety was that it "escapes the deficiencies or rigidity inherent in [either] the unitary or neat division of responsibilities model." Hence Kirlin advocated a flexible federalism, which he described as "a mixed system, in which different policies and intergovernmental fiscal mechanisms dictate quite different relationships among governmental units" (pp. 85-86).

11. A third policy type, referred to as allocative or allocational, provides services that tend to be viewed as neutral in their economic effect, the assumption being that they benefit the community as a whole. These include the various traditional areas of local government activity already established in the industrial city era, such as sanitation, public safety, and education, as well as government administration and other basic "housekeeping" duties. See Pagano and Moore (1985) and Peterson (1981).

12. The degree to which these partnerships have moved by now beyond the normatively sanctioned government/interest group relations generally associated with pluralist democracy and toward corporatist forms not traditional to the American political economy is a matter of some discussion. See, for example, Benson (1982) and Levine (1989).

13. See Keller (1991) for a vision of economic development in Baltimore driven by advances in biotechnology and the life sciences, such as medicine, biomedical engineering, and environmental technology. See also Bowie (1992) for a discussion of a number of possible obstacles to the full realization of that vision that are not unique to Baltimore. Besides uncertainties regarding capitalization and commercialization of promising research, these also include the serious weakness of the city's school system as a source of technically qualified workforce.

PART TWO

Evolution and Organization of the Urban Policy System

[4]

The New Deal

CREATING AN URBAN POLICY SYSTEM

To understand fully the reshaping of the basic forms of community settlement over the past half century, it is essential to recognize that American society has gone through several phases of institutional reorganization since the 1930s. A central aspect of this reorganization has consisted of changes in organizational relations between the different levels of government as well as between government and the private sector. My analysis of the changing stages of urban policy will examine both the changing objectives of urban policies over a number of major policy periods and changes in intergovernmental relations that have affected how the urban policy system has been organized to implement its declared objectives. Because policy is not just a matter of defining objectives but also of organizing and acting to achieve them, the second concern is as important as the first. Finally, I will examine also the growth of organizational relations between the urban policy system and related interests in the private sector.

In the present chapter, discussion begins with a brief review of the evolution of intergovernmental relations as a context for the emergence of an urban policy delivery system during the New Deal. New Deal programs are examined as key elements in the restructuring of the traditional intergovernmental system of federalism, contributing to the replacement of that system with a model stressing cooperative interaction between the federal and other levels of government. It was in the context of these changes in federalism that the foundations of a national urban policy system were first established, with implications for the structuring of urban programs for decades to come.

THE NEW DEAL: CHANGING
THE INTERGOVERNMENTAL CONTEXT

Since the economic and political crisis of the 1930s and the changes in governmental relations that it produced, urban policy has become increasingly a matter of intergovernmental activities involving units of federal or state government interacting with agencies of city government. By the 1980s it could be said justifiably that "there are . . . few urban functions—whether law enforcement, education, public health, transportation, or community revitalization—that do not involve all three levels of government" (Bollens & Schmandt, 1970, p. 148).

City governments still deliver these services, but many have come to depend significantly on federal and state governments for financial support. Here I refer particularly to the *intergovernmental city,* typically an older industrial city that has experienced serious economic decline and now depends on high levels of fiscal assistance from federal and state governments (Burchell et al., 1984; Kantor, 1988). In recent decades the capacity of many industrial era cities to deliver local public services has become dependent on the actions or inactions of the intergovernmental system in ways entirely unanticipated in the early part of this century. Thus, to comprehend fully the development of today's urban policies and programs, it is necessary to understand the continuing evolution of the intergovernmental system of which urban government has become a component element. With regard to this, two key questions can be identified. How have changes in the intergovernmental system as an interorganizational context for urban government affected the shaping and implementation of urban policy? How may they affect it in the foreseeable future?

The evolution of the intergovernmental system can be understood in terms of four stages of historical development: *Dual Federalism; Cooperative Federalism; Creative Federalism;* and the most recent stage, typically referred to as *New Federalism* (Reagan & Sanzone, 1981). I will describe briefly the first two stages here, discussing Creative and New Federalism respectively in later chapters.

Dual Federalism

This was the legally specified intergovernmental system of the United States until the third decade of the 20th century. Its roots can be traced to the entrenched values of governmental decentralization prevalent in rural America reaching back to colonial times, out of which emerged the concept of the independent sovereignty of the states at the founding of the Republic. Based on the dominant 19th-century interpretation of the Constitution, Dual Federalism emphasized a clear separation of the powers of the federal government from those of the states, with each regarded as properly acting only within the area of its defined sovereignty. In this perspective, all powers not explicitly designated as federal were reserved to the states. During the 19th century this meant that aside from a number of basic federal responsibilities such as regulation of interstate commerce and levying of federal taxes, virtually all domestic policies were funded and managed by the states alone (Bollens & Schmandt, 1970).

An important exception to this rule was the occasional grants of federally owned land to individual states, typically on condition that the recipient state use the land for purposes of economic or community development such as building railroads or erecting public schools. At times land grants simply were converted to cash by the state selling off parcels to help pay its debts. It was not until the close of the century that the first regular annual cash grant was provided to the states by the Hatch Act of 1887 to assist them in developing agricultural experiment stations. This was a modest step toward the later development of a full system of cash grants-in-aid involving federally defined objectives and program standards.

During the 19th century the entire grant-in-aid enterprise remained small, accounting for only 1% of federal expenditures by 1900 (Howitt, 1984, p. 4). Consequently, during this period federal influence on most state policies was still quite limited. With regard to policies related to urban development, there was no substantial federal involvement until the 1930s, even though the United States had been experiencing rapid urbanization and its related problems for over half a century and a majority of the national population was counted as urban by 1920.

During the course of the first three decades of the 20th century the constitutional barriers between federal and state governments began to erode as various functional relations between these two levels of government grew in response to the needs of an expanding, increasingly complex national society. The establishment of a federal income tax in 1913 provided an important source of new revenues with which to undertake early federal grant initiatives in areas such as highway construction, forestry, and agricultural extension service (typically to assist economic development in rural states), as well as vocational education and maternal and infant care in response to the needs of a rapidly expanding population (especially in major industrial cities). By 1930 cash grants to the states had grown to about 5% of total federal outlays, or some $200 million provided through a modest array of 15 grants, with highway construction the largest single grant (Howitt, 1984, p. 5). Yet despite these early 20th-century developments in federal-state grant relations, Dual Federalism—as a persisting legacy of earlier rural America—remained the predominant form of intergovernmental relations until the coming of the Great Depression.

Cooperative Federalism

In responding to the crisis of the Depression the American intergovernmental system changed decisively in both legal and functional terms. The "constitutional revolution" of the 1930s expressed a recognition of the necessity of shared responsibilities between federal and other levels of government, as against the earlier notion of separate federal and state sovereignties. Cooperative Federalism adopted the principle that two or more levels of government could share public functions (Reagan & Sanzone, 1981) because it recognized the usefulness of intergovernmental cooperation in addressing important public policy problems, beginning with the problems associated with the collapse of a predominantly urban-industrial economy.

Given the unreadiness of states to assist their localities and with one municipal treasury after another facing bankruptcy, it was only federal grants in support of work relief programs that made even moderately redistributive local programs possible.[1] In intergovernmental terms the "New Deal was a major turning point" because "[u]nder the pressure of a national economic crisis, the federal government vastly expanded the scope of its domestic programs, including aid to state and local governments" (Howitt, 1984, p. 5). The grants involved were typically *categorical grants* (also known as functional grants) by which the federal government contracted for specific, narrowly defined categories of services or functions to be performed by state or local governments. During the period from the mid-1930s to the mid-1970s, categorical grants became the major vehicle by which federal funds were transferred to other levels of government. In particular, during this period they came to serve as essential channels for the funding of urban policies through which federal aid flowed to local governments for purposes such as public works, urban renewal, and the programs of the War on Poverty.

Between 1933 and 1938 Congress doubled the number of federal grant programs to 31, including several large temporary emergency programs of public works. In dollar terms, the volume of federal aid increased about 10-fold over the level of 1930, reaching a peak of around $2.2 billion in 1935. By the late 1930s the typical federal expenditure for grants to subfederal governments had leveled off to about $1 billion annually, fully five times what it had been at the beginning of the decade. This represented a commitment of federal support for state and local physical and economic development that would have been inconceivable just a decade earlier, especially because much of it was concentrated in massive projects of public works repair and renewal in central cities, which as legal creations of the states had been treated traditionally as their exclusive jurisdiction and responsibility. New Deal programs in public works and public housing, together with federal mortgage insurance programs supporting private home ownership and construction in central cities and surrounding suburban areas, laid down the essential organizational foundations of a national urban policy system in the 1930s. Ever since that era, federal policies have had important implications for the redevelopment of central cities and for the challenges posed by suburban growth to central city economies and residential sectors.

NEW DEAL PROGRAMS:
IMPACTS ON URBAN POLICY AND DEVELOPMENT

Prior to the New Deal, the limited Cooperative Federalism that existed was essentially a matter of relations between Washington and the states. What little assistance went to cities was channeled through state governments, based on the traditional legal conception of cities as creations of their respective states. Direct federal-local relations were virtually nonexistent, first becoming established during the 1930s in order to implement emergency measures such as New Deal-financed programs in public works and housing (Martin, 1965). It was in this unexpected, unplanned-for context of national crisis that the early organizational foundations of an urban policy system were laid down. Proceeding on the basis of a process of political compromise and administrative trial-and-error, the New Deal period proved to be an important testing ground for programs of federal assistance to urban development. By the end of the period, this process had resulted in the creation not of highly centralized arrangements (contrary to much current rhetoric) but of significantly decentralized interorganizational models for federal program administration that have affected the implementation of major urban policies ever since.

In examining the development of the urban policy system during the New Deal and after, an organizational/policy perspective is particularly relevant because it points to the key issues involved in the relationship of government policy to the restructuring of American cities. Briefly, during any given period in the evolution of urban policy, there are two central issues to be resolved. First, what type of policy is being advanced? What are its objectives and content? Is it economic developmental policy supporting local economic growth, often largely to the benefit of established business and professional interests, or is it redistributive policy, which aims at reallocating resources to the community's less advantaged members?[2] Second, through what type of organizational arrangements—predominantly centralized or decentralized— are policy-implementing programs delivered to their intended targets? Ultimately this involves the structure of the particular intergovernmental policy system through which the program is delivered (e.g., urban, health, education) and is related to the stage of development of that system and the stability of established patterns of program delivery. Throughout this book I will examine the key issues of policy content and program delivery structure during several policy periods, relating each of these issues to the prevailing political context for urban policy as influenced by the balance between the major policy interests of the period.

Under the New Deal, Congress in the 1930s created an array of public agencies to deal with the crisis of the economy, including notably the crisis of the housing industry. The creation of these agencies occurred in a piecemeal experimental fashion; some were only temporary; others have endured to the present. Several were established to stimulate industry and provide employment through construction of public works, and others were set up to relieve the housing crisis. Together, four agencies came to form the organizational core of what may be conceived of broadly as the New Deal's improvised system of national urban policy (Bellush & Hausknecht, 1967).

Among the several public works and relief organizations created by the New Deal, two stand out by virtue of the organizational as well as physical infrastructure they left for subsequent urban development. These were the Public Works Administration (PWA) and the Works Progress Administration (WPA), created in 1933 and 1935, respectively. Of particular interest from this book's perspective, experience with the programs of these agencies led to the establishment of intergovernmental administrative relations that widely have structured the delivery of federal programs of importance to cities.

In the area of housing the public organizations at the center of the New Deal's urban-relevant policies included the following: with respect to private housing, the Federal Housing Administration (FHA), created in 1934; in the area of public housing, the United States Housing Authority (USHA), created in 1937. FHA's programs were developmental, created to stimulate the operation and development of the private housing market and targeted especially to middle-class home buyers. On the other hand, USHA's public housing programs were redistributive, intended particularly for the working poor whose housing needs were not being met by the private housing industry.

A review of the basic programs of these agencies will show how the two major issue variables—the type of policy output delivered by a particular program and the structure of the program delivery system—can be applied in analyzing the evolution of urban public policy. I will begin with a discussion of public works policies that influenced urban community development during the New Deal and left their mark on program structures as well as on physical facilities for decades afterward.

THE NEW DEAL AND PUBLIC WORKS

Of the public works agencies created by the New Deal, the Public Works Administration and the Works Progress Administration were the largest in terms of total expenditures and the most significant in terms of the constituencies that they brought under the New Deal umbrella. During the 1930s their public works programs gave a much-needed lift to the construction industry and won support from both labor and business interests. Although they were not intended primarily for the purpose of urban development, the facilities that these New Deal programs produced made a major contribution to the modernization of the urban physical plant, providing American cities with a variety of new public infrastructure ranging from school buildings and city halls to subway lines and high-speed highways (Mollenkopf, 1983).

Established in 1933, when almost 25% of the workforce was unemployed, PWA was a federal agency "designed to provide states and localities with financial assistance to aid them in constructing public capital facilities" (McKean & Taylor, 1955, p. 85). It specialized in supporting the construction of large durable public works (e.g., major bridges, dams, tunnels, and highways), typically employing mainly skilled labor for these technically complex projects. Its stated objective was to "stimulate industry through the purchase of materials" (notably steel and other heavy industrial products) "and the payment of wages" (p. 106). The underlying intention was of "priming the pump of business" with a variety of large-scale public works projects. Through a type of trickle-down logic prevalent during the early New Deal, the assumption was that the stimulus of public capital investment to basic sectors of the economy such as the steel and construction industries would diffuse its benefits throughout the economy and down to the broad working population. As a product of the early New Deal, PWA was a capital-intensive public works program with relatively indirect and delayed benefits for the unemployed.

In contrast, WPA was established in 1935 in response to pressure from newly organized interest groups (in particular, industrial labor and big-city officials) as well as to concerns within the Roosevelt administration that earlier programs as yet had failed to lift the country out of depression. WPA was devoted primarily to providing direct federal work relief as quickly as possible through a variety of labor-intensive public employment projects. Drawing on the experience of earlier New Deal emergency relief efforts, it provided for almost a decade the major public employment programs across a broad spectrum of the labor force from low-skilled to highly skilled (with emphasis on the former) and left its mark on numerous public works in cities across the country. Yet, it was PWA—the other major public works program—that became the long-term model for the administration of later urban programs such as public housing and urban renewal. To understand why, it is worth tracing briefly the administrative background to the creation of WPA.

WPA: Background and Operation

A major predecessor to the Works Progress Administration was the Federal Emergency Relief Administration (FERA), established in May 1933 to distribute matching grants to the states to assist state administration of cash relief to the unemployed. To receive these grants, states were expected to match each federal dollar with three of their own, and they were to administer the provision of relief payments themselves, preferably through professionally staffed state agencies (Robertson & Judd, 1989, p. 110). This essentially decentralized approach, delegating the main responsibility for programs of public relief to the states, presented some serious problems. In particular, there were wide variations in the financial and administrative capacity of different states to respond to FERA grant requirements. Although some larger, wealthier states (such as New York, for example) were able to match fully the federal funding, others could match only partly, and some did not match at all but simply reduced local relief funds and substituted federal dollars received through the grants (Jansson, 1988). In the worst cases, misuse of funds for political patronage or outright corruption became sufficiently serious that

FERA administrators federalized the programs of half a dozen states, operating them with FERA personnel (Robertson & Judd, 1989). Another drawback of FERA was that it provided only cash relief but not work to the unemployed. This ran counter to Roosevelt's preference for work relief as a means of preserving the recipient's morale and self-respect. Eventually, the FERA program was phased out at the end of 1935.[3]

In contrast to FERA, WPA was a redistributive agency that focused on programs of work relief and "directly employed workers through a network of local offices . . . to work on labor-intensive local public works projects" (Mollenkopf, 1983, p. 66). Again unlike FERA, WPA was federally centralized, its chief administrator being given "more extensive authority with respect to the distribution of public funds and the control of local activities than any other governmental officer in time of peace" (Magill, 1979, p. 58). Over its working life, WPA's own experience demonstrated the suitability of centralized administrative structure for implementing redistributive federal objectives at the local level in a speedy and effective manner, particularly when state officials do not have the same commitment to local redistribution as do federal policy makers.[4]

Although the basic goals of WPA were redistributive, its work relief programs also made a notable contribution to the public infrastructure of American cities. During the period from 1936 to 1940, it put to work an average of more than 2.3 million workers each year (whereas employment in PWA was only about a half million at its peak in 1934). WPA's major activities included highway and road improvements, as well as construction and maintenance of water and sewage systems, public buildings, parks, and recreational facilities. Its projects were concentrated largely in urban centers, where it built tens of thousands of miles of new streets and sidewalks, as well as thousands of bridges and viaducts. Among other achievements, it "built or improved more than 2,500 hospitals, 5,900 school buildings, 1,000 airport landing fields, and nearly 13,000 playgrounds" (Advisory Commission on Intergovernmental Relations [ACIR], 1982, p. 16; Chandler, 1970).

In addition, WPA assumed responsibility not only for manual work programs of this type but also for programs such as those of the FERA's Civil Works Service Program and Emergency Education Program, which provided work for a wide variety of service occupations, including artists, painters, writers, actors, musicians, historians, and archaeologists, as well as teachers, clerical workers, nurses, and nutritionists (Chandler, 1970).

This broad array of federal public work relief programs for the unemployed apparently disturbed many conservatives, who saw WPA's centrally managed employment activities as a threat to the operations of the private labor market; hence they vigorously attacked the record of WPA accomplishments, picturing it as little more than make-work of questionable productivity (ACIR, 1982). This negative image was sufficiently persuasive that despite the actual record, WPA projects continued in later years to be frequently stereotyped as having amounted essentially to boondoggles or "leaf raking." Ultimately, although this federal agency made major contributions to providing employment and modernizing public infrastructure, it essentially was rejected as a model for later federal program administration.

From a current perspective, a twofold observation is in order here. In political terms, during the Depression it became permissible for the federal government to intervene locally with programs such as WPA that provided emergency employment (thus quickly increasing local economic resources) even if those programs were controlled centrally, so severe was the economic crisis. On the other hand, such allowances would not be made for programs that did not directly provide employment benefits to the locality. Those programs would have to be structured not to encroach on either established local business prerogatives or traditional states' rights—in short, structured in a basically decentralized manner. In interorganizational terms, member units of the local organizational field tend to resist the restructuring of their existing organizational domains when faced with the addition of new organizations (or organizational programs) that derive from outside the local field (Warren, Rose, & Bergunder, 1974). Such is the case unless externally originated and controlled programs such as WPA provide resources crucial to the continued survival of existing organizations, or externally originated programs are made sufficiently subject to local field control to assure that they will not encroach significantly on the domains of the major established field units (see the case of PWA that follows).

PWA: Operation and Implications

In contrast to WPA, the Public Works Administration emphasized the building of large, technically complex facilities such as New York's Lincoln Tunnel and the Grand Coulee Dam, as well as a range of other large public works. During the first 6 years of the Roosevelt administration, PWA spent almost $5 billion for the construction of a wide variety of large-scale infrastructure, such as highways, dams, major bridges, water and sewage systems, and airport facilities, as well as public buildings including courthouses and city halls. Although these various facilities were deployed throughout the nation, they had particular significance for the modernization of urban infrastructure, with over half of PWA expenditures going to the construction of major public facilities in urban areas (Huthmacher, 1968, p. 130).

Although employment in PWA's projects provided some work relief, it was considerably less than that provided by WPA because PWA projects were generally more capital intensive. At its peak in November 1938, WPA employed just over 3.3 million workers; by contrast, peak employment for PWA was 541,000 in July 1934, or about one sixth that for WPA. Thus total expenditures over the duration of these programs were $13.4 billion for WPA and $4.5 billion for PWA (ACIR, 1982, p. 14). Although WPA and PWA both had developmental results, WPA's work relief mission focused its efforts on the primarily redistributive objective of providing jobs and wages to the unemployed. On the other hand, the main objective for PWA was cost-effective production of major public infrastructure; its operations were essentially developmental.

Unlike WPA, which was a federally centralized agency, PWA operated through relatively decentralized intergovernmental relations. Organizationally, PWA established ties with thousands of cities and counties, as well as special districts, to which it made grants and loans. It typically did so by working with a local public authority that could legally exert the power of eminent domain to acquire property and could borrow money if necessary. No less important, this authority could recruit a staff that would work independently of local political machines. PWA actively structured these organizational arrangements through proposals for enabling legislation that it addressed to state legislatures across the country, recommending creation of local authorities to implement its programs. These proposals generally received political support from a wide array of urban interests, including businesspeople, contractors, attorneys, and construction unions, all of whom stood to benefit from the local contracting out of PWA projects (Mollenkopf, 1983).

Thus, by delegating the administration of its programs to local public authorities, PWA successfully mobilized local area interests on its behalf. This enabled it to get around the traditional urban machines and build its own supportive political coalitions. These new local development coalitions brought together formerly disparate political elements in the community—particularly business, organized labor, and public officials—who were ready to cooperate in expanding local employment through public infrastructure development (Mollenkopf, 1983). PWA's approach, in contrast to that of WPA, was to provide a broad distribution of benefits to a variety of interests in the community, rather than targeting them specifically to unemployed workers. Through its basically decentralized structure of program delivery, PWA was more readily received than WPA within the multiple-interest environment of city politics, particularly because its economic development objectives appealed to the business sector, always an influential set of actors in the urban policy arena.

Over the long term, PWA's distinctive structuring of federal-local program relations contributed to shaping the "institutional groundwork" for the participation of economic growth interests that have played the influential role of "postwar progrowth coalitions" in much of later urban politics and policy making (Mollenkopf, 1983, p. 66). The federally decentralized administrative approach first associated with PWA subsequently has been found suited to the multi-interest distributive politics of urban redevelopment and even has been put to use as the delivery model for such explicitly redistributive programs as public housing.

THE NEW DEAL AND PUBLIC HOUSING

Before discussing the evolution of public housing policy under the New Deal, a brief refresher is useful. In the general area of housing, the public agencies at the center of the New Deal's urban-relevant policies included the following: the Federal Housing Administration with respect to private housing and the United States Housing Authority in the area of public housing (Bellush & Hausknecht, 1967). FHA's programs were developmental, aimed at stimulating the development of the private housing market and targeted to middle-class home buyers. On the other hand, USHA's public housing programs were redistributive, intended to provide access to affordable housing for the working poor. As two very different types of housing policy, the different interests and objectives they represent have been frequently in contention.

At the outset here, it should be noted that the primary thrust of New Deal national housing programs was not connected with public housing. The New Deal's most important housing program was established as early as 1934 with the creation of the Federal Housing Administration, authorized to provide mortgage insurance for private housing. Over the long term, FHA's mortgage program played a major role in encouraging construction of private housing and in the rapid development of the nation's suburbs. It was highly popular not only with private homeowners and builders but also with the banking and home loan industries that it insured against the financial risks of mortgage lending. Based on such broad and influential support, the objectives of the federal mortgage program came to be accepted far more readily—and to be more important to post-Depression processes of change in and around America's central cities—than any subsequent federal initiatives for public housing, which were typically overshadowed by the scope and impact of FHA's operations.

In contrast to the early establishment of FHA, the Roosevelt administration waited until 1937, after having won a second term and a certain degree of political security, before creating the United States Housing Authority to administer a public housing program targeted to the housing needs of lower income groups and the "Depression poor." The main objective of the program was to redistribute resources to the working and unemployed poor to help ameliorate their living conditions. The earlier experience of the WPA program suggested that such redistributive policy would be implemented most appropriately through a centralized program delivery structure, either by direct federal program administration or by closely monitored federally regulated intergovernmental programs. Yet despite its explicitly redistributive objectives, the administration of the public housing program came to be patterned in accord with a relatively decentralized intergovernmental model, which subsequently served as the program delivery model not only for public housing but for the urban renewal program as well. This can be understand better by examining the early evolution of New Deal public housing policy.

Public Housing Decentralization
and Creation of "Local Authorities"

The adoption of an intergovernmentally decentralized administrative model for public housing can be traced to the New Deal's Public Works Administration. Under the National Industrial Recovery Act of 1933, this federal public works agency was given authority to engage in the clearing of slums and their replacement by low-rent public housing (Bellush & Hausknecht, 1967). PWA soon established a Housing Division, which at first tried to induce private developers to construct low-cost housing by offering them reduced-interest loans if they agreed to voluntarily limit their profits.

But there was little response from the housing industry to this proposal; hence PWA undertook to directly acquire and develop project sites at the local level by exercising the power of eminent domain. Under this approach, PWA's public housing program thereafter would be centralized administratively at the federal level. However, structured that way, the program soon provoked vigorous political and legal opposition to the federal application of eminent domain in local municipalities. In spite of having established a record of speedy, decisive action in providing low-cost housing, in 1935 PWA effectively was forced to give up its housing construction activities following a successful court challenge in Louisville, Kentucky, to its proposed use of eminent domain in local slum clearance (Hays, 1985).

In those circumstances, the Roosevelt administration abandoned the notion of a centralized program involving direct federal action to construct public housing locally. Instead, the approach now adopted was one of decentralized governmental action through specially created *local authorities*.[5] Here the Public Works Administration turned to a modified form of the local authority mechanism originally developed to administer construction of its public works projects at the local level. PWA took a step beyond those earlier arrangements by encouraging states to pass legislation that would allow the establishment of *local housing authorities* (LHAs) to which it then legally could provide funding not only to implement construction but also to manage long-term operation of public housing projects. What was decisive in winning acceptance of this approach was the provision that the LHA would be "an independent local government agency whose officers would be appointed by the mayor or city council" (Henig, 1985, p. 163), rather than merely a local extension of PWA. Instead of operating directly as a branch of the federal works agency, each local housing authority would be an intermediary quasi-public organization, linking federal program objectives with interested local constituencies such as real estate agents and building contractors.

Since that time, LHA officers typically have been selected from leading private segments of the local community, such as banking, real estate, legal, and other local business or professional interests. In this is seen the meeting of public and private sectors in a specially organized intergovernmental administrative structure, established to incorporate private interests as partners in the local implementation of federally subsidized housing policies.

The Roosevelt administration now focused on persuading states to adopt legislation to allow the establishment of local housing authorities empowered to carry out public housing programs. Within 2 years, almost 50 such authorities had been created in 29 states (Robertson & Judd, 1989, p. 305). In this way PWA's decentralized model of administration, originally associated with basically developmental public works projects, was generalized to the public housing program, even though the housing program had been established to serve objectives that were clearly redistributive. Thus the public housing program's intergovernmental delivery structure was shaped to bypass legal-political obstacles imposed by lingering traditions of Dual Federalism and the vigorous opposition of congressional conservatives, rather than as the most effective organizational means by which to implement redistributive objectives in housing.

PWA's delegation downward of federal program authority to local public housing agencies became the guiding framework for the administration of public housing policy in the United States and later influenced the structuring of the urban renewal program as well.[6] This framework was soon adopted in the Wagner-Steagall Low Rent Housing Act of 1937, which created the United States Housing Authority and provided for a national public housing program in which housing projects would be constructed, owned, and managed by local housing authorities. The federal government, having delegated part of its authority to these local agencies, then enters into a set of contractual arrangements with each agency. In these contractual relations, Washington accepts certain restrictions on its own authority to create low-cost housing. Under the Housing Act of 1937, the federal housing authority cannot directly select public housing sites; it only can approve or reject sites selected by the local housing authority. In addition, the acquisition of construction sites and their development, administration, and ownership are left in the hands of the local agency (Bellush & Hausknecht, 1967). The USHA would contract with the local housing authority "to pay the annual principal and interest on long-term, tax-exempt bonds, which finance construction" under the supervision of the local authority (U.S. Department of Housing and Urban Development [U.S. DHUD], 1974, p. 9). Thus, instead of housing projects being constructed directly by the federal government, Washington "would supply the funds and overall direction, but administration of the public housing program would be left to the localities, through local housing authorities . . . created and empowered by state enabling acts" (Hartman, 1975, p. 114).

Consequences of Decentralization

By the mid-1940s such delegation of federal administrative authority over public housing to local agencies established a context for long-term continuing conflict in the public housing program:

> This devolution of authority . . . set the stage for community battles over the siting of public housing projects which became a recurring feature of local politics over the next 40 years. . . . The enabling legislation in many states required that local participation in the program be subject to direct voter approval by referendum.

These referenda gave opponents the opportunity to excite public fears. (Hays, 1985, p. 90)

Following World War II, in one community after another the state-required referendum process encouraged opponents to mobilize opinion against public housing, often by using racial scare tactics, and to exert pressure on local politicians to confine public housing sites to poor and minority neighborhoods. As a consequence, through its restriction to areas already viewed as slums, public housing became stigmatized in the opinion of many urban residents.

This was not the only drawback of decentralized administration, or as some refer to it, *localization* of the public housing program. Critics of localization see it as a basic weakness of design in a program with redistributive goals. As Hartman (1975) observed:

All basic decisions about public housing—whether to have any units at all, how many to have, where they should be located, what type to have—are placed in the hands of an appointive local body that historically has tended to represent those elements in the community—businessmen, lawyers, and realtors—that were opposed to or ambivalent about extensive government subsidies for housing the poor. (p. 114)

The decentralized structure of the public housing program has affected deeply its implementation in that it leaves it vulnerable to unfriendly local interests who have been allowed to define its image by determining its location, scale, and layout in the community. Localization has affected decision making on a wide range of substantive programmatic issues, producing severe limitations on program operation in areas such as client eligibility and program finances as well as on possibilities for innovative project design and management. The overall result has been a narrowing of the constituency for public housing and its widespread loss of legitimacy in public opinion.

Critics of decentralization have noted that LHAs, with their essentially middle-class membership, generally have failed to advocate significant improvements or enlargements in local public housing programs. In addition, for those projects that they have endorsed, they have tended repeatedly to select sites that are socially segregated and often physically marginal in relatively undesirable locations away from central facilities. In the exceptional cases where LHAs actively supported provision of more public housing in the community, state requirements that public housing construction be subject to approval by public referendum have served as important obstacles. Perhaps most serious has been the outright exclusion evidenced particularly in smaller cities and rural small towns that have refused to establish housing authorities at all. In consequence, as of the mid-1970s over 55% of the national population was unserved by local public housing authorities (Hartman, 1975, p. 115), effectively leaving the task of providing affordable housing for low-income groups to larger, generally older cities.

Changes in Federal Policy Toward Public Housing

Another major obstacle to the growth of the public housing program since its inception in 1937 has been the frequent unwillingness of Congress to authorize sufficient units to meet the backlog of need or to appropriate full funding for the construction of units already authorized. By 1939 the USHA had committed $800 million in project loans to local authorities, as a result of which just under 170,000 public housing units were constructed, mostly in a few large central cities. At that point, a coalition of southern Democrats and Republicans joined together in the House to block any further authorizations. This conservative coalition, with heavily rural and later increasingly suburban constituencies, repeatedly voted to minimize or reduce public housing allocations over the next several decades (Mollenkopf, 1983, pp. 69, 79, 117). The coalition also has appealed often to considerations of local or states' rights as a basis for maintaining decentralized administration of public housing.

With the coming of World War II the prevailing stalemate regarding public housing was suspended, and a large number of publicly subsidized units was constructed in a relatively short time. This was done under the auspices of the National Housing Agency, created in 1942 to centralize all federal housing authorities under a single agency for the duration of the war. Given the context of military emergency and the need to expedite the provision of defense-related housing, localistic concerns were set aside and the federal government proceeded with direct construction of over 850,000 units of housing for military personnel and defense workers under the statutory authority of the Lanham Act of 1940 and subsequent wartime legislation. In the space of 4 years this public housing effort produced approximately five times as much housing as the civilian public housing program of the preceding 2 years. After this record-breaking achievement, "lacking the stimulus of the war effort, the Federal Government abandoned its role of directly supplying housing"; not only did it give up the role of direct, efficient provider of publicly needed housing, it then "demolished two-thirds of the wartime-constructed units and sold the remainder" (U.S. DHUD, 1974, p. 10).

It was not until the Housing Act of 1949 that legislation authorizing a major increase in public housing units was passed. The act authorized construction of 135,000 units annually for 6 years. This would have provided 810,000 units by 1955, almost as many as those demolished or privatized after the war. However, Congress consistently failed to fund the program to the level authorized, so that actual appropriations covered between a peak of 90,000 units in fiscal year 1950 and zero in 1954. By 1952, only 85,000 units, or a little more than 10% of those authorized, were under construction, and "by 1960, five years after the target date for completion of the 810,000 new units, less than one-quarter of these had been built" (Hays, 1985, p. 93). As late as 1965, a total of only 605,000 units of public housing were in existence, or less than three fourths of the total new units authorized in 1949 for completion by 1955. After that, approximately another 675,000 were added over the next 15 years, fully 85% of those being constructed during the decade from 1965 to 1975 (U.S. Bureau of the Census, 1981, p. 768).

Beginning with the Nixon administration in the early 1970s, there has been a long-term tendency for presidential administrations to minimize requests for authorizations of additional units. Public housing construction came to a halt after the declaration of a moratorium on it in January 1973, following Nixon's landslide reelection. New construction was revived briefly on a reduced basis under the Carter administration, but then was phased out entirely by 1987 under the Reagan administration (Hays, 1985). By the early 1990s, with a national poverty population officially counted at over 30 million, less than 2 million units of public housing currently serve all eligible households, with waiting lists of at least several years in major cities.

From the initial promise of the public works and slum clearance programs of the New Deal, which first opened the prospect of large-scale modernization and redevelopment of older American cities (with a particular focus on older inner-city residential areas), we now have come to the stigmatization and stagnation of contemporary public housing. Clearly, public housing has not held high priority with government decision makers, and its consequent neglect is reflected not only in the poor physical condition of much public housing but also in its typical confinement to areas of segregated and seriously deteriorated housing in the cities where it has been concentrated. Over the past half century, this lack of official commitment to public housing is striking in comparison with strong and sustained governmental support for construction of private housing, especially that associated with suburban development. The organizational/policy context for private residential construction, as established under the New Deal, is discussed in the next section.

THE NEW DEAL AND PRIVATE HOME OWNERSHIP

As I have observed, in intergovernmental affairs the New Deal represented a significant departure from the traditional pattern of Dual Federalism that had tended to minimize involvement by the federal government in local domestic matters. The New Deal not only took major steps toward a more interactive type of federalism but also produced important new relations between government and the private sector, reflected in public policies aimed at reviving the economy. In the early 1930s it did so with regard to the housing industry by extending public support to the operations of the private housing market, providing mortgage loan insurance, and encouraging easy credit to help revive the home finance and construction sectors.

Origins of a Housing Program for Homeowners

These measures for economic recovery were associated with the creation of new public organizations among which the foremost in the area of private housing was the Federal Housing Administration. Through such agencies the federal government acted to assist the organizations of the construction and banking industries by underwriting their economic risks, thus protecting the future profitability of their housing-related operations. Such policy can be conceived of as progrowth or developmental policy. Besides assisting the growth of the home-building industry and the banking and loan organizations that provide it credit, another important effect of such developmental policy has been to encourage the growth of middle-class home ownership, in the process contributing to the rapid spread of post-World War II suburban development with significant impact on central cities.

Organizationally, FHA's mortgage program was structured to enable the agency to support development of private housing by providing decentralized incentives to the banking industry, rather than itself undertaking the operations of the industry through a centralized federal program. Although the FHA's mortgage insurance program did not involve intergovernmental relations, it entailed multilevel interorganizational linkages that connected the federal agency to the networks of home financing organizations at the state and local levels across the nation. It placed the organizational and financial resources of the federal government at the disposal of banks, insurance companies, saving and loan associations, and related financial organizations providing mortgage loan capital to private home buyers in their local communities. Thus, although the New Deal often has been pictured as an era in which public policy was devoted predominantly to creating federally central-ized programs for redistributive purposes, the major response of federal policy makers to the crisis of the housing industry—FHA's mortgage insur-ance program—can be shown to have been significantly different from this image.

Reacting to the massive occurrence of mortgage default during the early 1930s—when nearly 1,000 mortgage foreclosures occurred daily and new mortgage lending and home construction went into deep decline—Congress established the Federal Home Loan Bank System, the Federal Savings and Loan Insurance Corporation, and the Federal National Mortgage Association, as well as the Federal Housing Administration (U.S. DHUD, 1974, pp. 7-8). The immediate objective was "to stimulate the private sector to build housing and to help individuals retain their homes or acquire new housing" (p. 8). Yet on a long-term basis, by creating these federal agencies to support the opera-tions of the home financing industry, federal policy decisively altered the nature of housing credit markets. Each of these public agencies became elements of the nation's organizational framework for home purchase and ownership, together exerting a major influence over the operations of the American housing industry and through it over basic social-spatial processes of change such as urban deconcentration and suburban development.

The first major New Deal housing legislation, the Federal Home Loan Bank Act of 1932, established a network of regional home loan banks and a Federal Home Loan Bank Board authorized to advance funds to participating banks and other financial institutions in order to secure home mortgages. The Home Loan Bank Act was followed by the Home Owners Loan Act of 1933, an emergency measure creating the Home Owners' Loan Corporation (HOLC). This agency loaned $3 billion to refinance a million defaulted home mortgage loans by way of direct low-interest government loans to affected homeowners. HOLC's loan program assisted 20% of all mortgaged urban homes (Mollenkopf, 1983, p. 70) and rescued the residential mortgage system from collapse. At the same time that this measure helped savings banks survive the financial crisis, it also saved thousands of individual homeowners from foreclosure and thus attracted an important element of the middle class as a loyal constituency for the New Deal. The act also established the early basis for the system of federally chartered savings and loan associations and empowered the Home Loan Bank Board to regulate that system.

Following close on the heels of HOLC, which it replaced, the Federal Housing Administration was created under the National Housing Act of 1934, partly in connection with attempts to revive the construction industry and assist the general recovery of the economy. The official rationale for creation of FHA emphasized that it would provide aid to families exposed to mortgage foreclosures, but there was more at stake than that. Widespread inability to meet mortgage obligations meant a severe weakening of the market for housing, increasing the loss of jobs in the building trades and adding to the overall level of unemployment (Hays, 1985). It also posed a serious economic threat to sectors of the banking industry involved in providing home mortgage loans, particularly savings and loan associations and other local community banks. Thus the basis for creating FHA in 1934 went considerably beyond HOLC's earlier aim of extending emergency loan aid to home-owning families.

FHA and Mortgage Insurance

Taking these considerations into account during the year after it had created HOLC, Congress decided to replace that agency's program of direct loans to homeowners by a program of federal insurance to mortgage lenders. The Housing Act of 1934 established federal support of the housing credit system through the provision of mortgage insurance by FHA to banks and other lending institutions. The objective was to arrange "a fundamental restructuring of the way people borrowed money for home purchases." By institutionalizing the long-term, low-down-payment mortgages first introduced by HOLC, "the FHA program greatly broadened the segment of the U.S. population which could afford to purchase a home" (Hays, 1985, p. 81). In particular, families of moderate income in the upper working and lower middle classes—overall, the economically dislocated middle class of the Depression era—were able now to afford home mortgages.

In short, to encourage revival of the residential construction industry during the Depression, the federal government provided insurance for mortgage loans made by private lenders, thus protecting them from the problem of default. The costs of this insurance are covered not by the banks or other lenders but by a fund accumulated from monthly fees paid by the borrowers whose loans are insured. However, FHA does not insure the borrower; in case of default on mortgage payments, the borrower loses the property. The agency insures the local lender against the costs of default, so long as the terms of the loan and the characteristics of the borrower satisfy FHA criteria. Further, it is left to the discretion of the local lending institution to qualify the borrower, that is, to decide whether FHA's criteria are met by the applicant for the loan. One very important consequence of this arrangement is that unlike the

HOLC, which directly insured the homeowner's mortgage, FHA placed the implementation of the federal mortgage insurance policy in the hands of local private financial institutions. This type of public-private devolution of authority permanently decentralized the federal mortgage insurance program and incorporated into national housing policy the perspectives of the bankers and mortgage lenders who participate in its local implementation.

The underlying objective was to induce banks to invest in home mortgages by shifting to the government the financial risk involved should the home buyer default on mortgage payments (Nader Congress Project, 1975). At the same time the idea was to stimulate broad middle-class family investment in housing with affordable loan terms for borrowers. In return for federally underwritten mortgage insurance, participating banks had to liberalize their loan terms. This meant reducing mortgage interest rates from a typical 10% or more to a maximum of about 5%, reducing down payments, and extending the loan repayment period from the prevailing maximum of about 10 years to a new maximum of 20 years. FHA mortgage insurance made possible the modern low-down-payment long-term mortgage of 20 to 30 years maturity. Over the extended mortgage term, monthly payments now could pay off the full principal gradually, rather than leaving a large final payment that often resulted in default (U.S. DHUD, 1974, p. 8).

Despite the inducements to mortgage lenders embodied in FHA's mortgage insurance program, bankers complained that long-term mortgage loans tied up investors' money, preventing quick conversion back to cash when more profitable investment opportunities occurred. In response, the 1934 Housing Act also authorized creation of the Federal National Mortgage Association (FNMA), which could raise funds by issuing government notes and bonds. This provided a secondary market for FHA-insured mortgages, "where approved lending institutions could unload their FHA-guaranteed mortgages in return for cash raised by FNMA with government backing" (Nader Congress Project, 1975, p. 156).[7] Finally, as part of its efforts to support the savings and loan industry, the act created the Federal Savings and Loan Insurance Corporation to insure savings accounts.

In summary, what did New Deal-created housing agencies and their programs for the nonpoor accomplish? Briefly, they assured an adequate flow of housing credit, saving millions of homeowners from mortgage foreclosure and making possible home ownership as a goal for a majority of the population. This had important implications for national political as well as economic stability and for the social-economic security of the upper working class and middle class in the wake of the severe crisis of the Great Depression.

The heart of the system was the federal government's decision to reduce economic risks to banks and related financial institutions through the underwriting of those institutions' mortgage loans to a growing home-owning population. These inducements encouraged the flow of real estate capital to the financing of homes for the broad upper working and middle class, with borrowers effectively paying for the costs of the program through insurance premiums attached by the banks to the mortgage loans they granted.

However, the FHA's mortgage insurance program did not address the housing problems of the lower working class or the minority poor. This tilt in the FHA's mortgage program in favor of serving the housing needs of predominantly White midscale families became even more pronounced in the postwar period. Even though mortgage insurance programs expanded to help the housing industry cope with rapid growth in demand following World War II, Blacks continued to be bypassed for loans by the agency. Over the long run, FHA's refusal to underwrite mortgages in inner-city neighborhoods contributed significantly to their rapid decline in the postwar period (Hays, 1985).

As discussion in this chapter indicates, FHA has proved to be the most important program created under the New Deal with regard to overall impact on the built environment, particularly in metropolitan areas. Yet despite its being the most significant urban-relevant program of a supposedly redistributive era, FHA presents a prime example of a program whose objectives throughout have been clearly developmental rather than redistributive. As late as the mid-1960s and the advent of the antipoverty programs, less than 3.5% of purchasers of FHA homes could be classified as poor or near poor; the overwhelming proportion, amounting to almost 85%, ranged from lower to upper middle class (National Commission on Urban Problems, 1968, p. 100).

Over the long term, besides making home ownership widely available to the nation's middle class, FHA's major contribution has been the part it has played in fostering suburban development and the departure of much of the urban middle class from central cities after World War II. In short, FHA has acted as "a vital factor in financing and promoting the exodus from the central cities and in helping to build up the suburbs. That is where the vast majority of FHA-insured homes have been built. The suburbs could not have expanded as they have during the postwar years without FHA" (National Commission on Urban Problems, 1968, p. 99).

Discussion of the impact of FHA programs on suburban development will be continued in Chapter 5. In concluding the present chapter I review the key aspects of the development of urban policy coalitions from the New Deal to the 1970s, prior to the establishment of New Federalism.

THE EMERGENCE OF URBAN POLICY COALITIONS

In this chapter I have described the beginnings of a national urban policy system. Here I discuss the interest groups that composed the emerging urban lobbies related to this system and the key points of their development to the early 1970s. Initially, at the core of the urban policy system stood those agencies in the executive branch of the federal government concerned with public works, public housing, and private home mortgage insurance and markets: PWA, WPA, USHA, FHA (the "big four"), and FNMA. Other organizations in the executive branch to which those agencies and their related lobbies looked for support or for some signal of the incumbent administration's policy priorities and intentions were the White House (the president and his domestic policy advisory staffs), the Budget Bureau (later the Office of Management and Budget), and the Departments of Treasury, Commerce, and Labor (Fox, 1972). Other relevant agencies added in the postwar period include the Department of Health, Education, and Welfare (HEW; later divided into the Department of Health and Human Services and the Department of Education) and the Department of Transportation.

During World War II, which finally brought full employment, the public works agencies (PWA and WPA) were phased out. In 1947 FHA, the Public Housing Administration (PHA, USHA's successor), and the Home Loan Bank Board were brought together as basic elements of the Housing and Home Finance Agency (HHFA) (Jacobs, Harney, Edson, & Lane, 1982, p. 10). Subsequently, in 1965 FHA and PHA were incorporated into a new cabinet department whose mission was broader than just housing—the Department of Housing and Urban Development (HUD). Outside the executive branch, in the Congress the most important actors with reference to the new urban policy system included the House and Senate banking committees (the major standing committees concerned with urban legislation), the appropriations committees (with jurisdiction over the appropriation of funds to federal programs), and the committees on government operations with their subcommittees on intergovernmental relations (Fox, 1972).

In general, organizations formally outside the federal urban policy system yet with close ties to it have constituted an organizational environment of competing groups and coalitions to which this policy system relates its decisions and actions. This organizational environment includes a variety of interest groups, which can be divided broadly between an *urban lobby* and an often contesting lobby of generally conservative opponents to urban-specific or closely urban-related policy and its associated program expenditures. Since the creation of HUD in the mid-1960s, the urban lobby generally has consisted of the United States Conference of Mayors (USCM), the National League of Cities (NLC), the National Housing Conference (NHC), the National Association of Housing and Redevelopment Officials (NAHRO), the National Association of Home Builders (NAHB), and during the 1960s the Urban Coalition, prominently including urban corporate business leaders. These organizations "often ally themselves with sympathetic groups from other functional areas," including labor organizations, racial or ethnic lobby groups, and urban planners (Fox, 1972, p. 103).

Opposed is the so-called *antiurban lobby*. This organizational coalition has consisted essentially of three major interest blocs: banking associations such as the American Bankers Association, Mortgage Bankers of America, U.S. Savings and Loan League, and the National Association of Mutual Savings Banks; business associations such as the National Association of Real Estate Boards (NAREB), the National Association of Home Builders (until the 1960s), U.S. Chamber of Commerce, and National Association of Manufacturers; and organizations of nonurban government officials such as the Council of State Governments, National Governors' Conference, and the National Association of Counties.

Overall, the urban lobby has consisted of groups that are themselves either composed of public officials (USCM, NLC, NAHRO) interested in a changing mix of public works, public housing, and urban renewal, or of public officials and citizen advocates associated with programs for subsidized housing (National Housing Conference) or low-to-moderate income neighborhood development (National Association of Neighborhoods). The opposing lobby has tended to consist of groups representing private interests in banking and real estate (NAREB, Chamber of Commerce, U.S. Savings and Loan League, Mortgage Bankers of America) associated with development of private housing and of state or county public officials interested in suburban housing and community development.

Organizationally, the two major milestones for the period 1933 to 1965 were the creation of the big four agencies of the 1930s (the early core of an urban policy system) and the establishment of HUD in 1965. Between these two the most significant organizational event for the urban policy system was the creation by President Truman of the Housing and Home Finance Agency in 1947 through executive reorganization. This produced a rather loose combination of agencies within HHFA: on the one hand FHA and the Federal Home Loan Bank Board, supported strongly by the real estate segment of the antiurban lobby; on the other hand, the Public Housing Administration, supported by the public housing segment of the urban lobby. Although these agencies were far apart in terms of their self-designated missions and philosophies, it was Truman's intention to produce a single agency that would be responsible for conducting a comprehensive housing policy. Perhaps with strong leadership dedicated to that goal, it might have been possible. However, facing a conservative majority in the Congress, he was able to gain approval for the new organization only by appointing as its head Raymond Foley, a former FHA administrator who was openly known to lack enthusiasm for public housing.

Within a few years, by virtue of the Housing Act of 1949, urban renewal had been added to HHFA's combined missions. HHFA now included the FHA (private housing), the URA (Urban Renewal Administration), the PHA (public housing), and several other disparate components, "each of which nurtured a set of supportive relationships with outside client groups and lobbied against the others for a larger share of whatever urban or housing-related monies Congress might appropriate" (Nader Congress Project, 1975, p. 227). Of all these, FHA traditionally was favored with the decisive support of powerful private interests such as housing builders and mortgage lenders, each with extensive connections to leading members of the congressional banking committees. On the other hand, the public housing and urban renewal agencies in HHFA (and later HUD) had to look elsewhere for support and thus developed relationships with public interest groups or professional groups such as the American Society of Planning Officials and the American Institute of Planners, with northern urban congressional liberals, and with quasi-public professional organizations such as the National Association of Housing and Redevelopment Officials, which evolved from the National Association of Housing Officials when the urban redevelopment program was created in 1949.

Because of the power and resources available to the Federal Housing Administration based on its special relations to banking and real estate interests (and dwarfing those of the public housing administration), the essentially suburban growth policy orientation of FHA prevailed at HHFA and then HUD until well into the 1960s. Thus FHA can be viewed as the executive agency anchor for the antiurban coalition, which until the 1960s remained fairly stable. That coalition justly could be thought of as the antiurban lobby until one of the major elements of the coalition, the National Association of Home Builders (representing small- and mid-sized home builders), emerged during the mid-1960s as an active supporter of federally subsidized housing programs. This occurred largely because of home builders' recognition that in an increasingly inflationary period "the conventional FHA mechanism was losing its punch and that stronger medicine would be needed to prime the homebuilding pump in the tight-money period of the 1960s and 1970s." By the early 1970s, NAHB was followed by several other lobbies for subsidized residential construction, including the Home Manufacturers Association and the Council of Housing Producers, representing large corporate interests seeking subsidies for industrially mass-produced housing, as well as liberalization of building codes for such housing (Nader Congress Project, 1975, pp. 211, 217). Thus, over time, the urban and antiurban (today called more appropriately *suburban*) coalitions have been distinct yet somewhat flexible, changing their composition in relation to changing circumstances of the economy and to changing opportunities offered by government programs, particularly those of the federal government.

NOTES

1. Overall, urban-relevant programs of the Depression era were not primarily redistributive. During the New Deal phase of Cooperative Federalism, federal policy choices continued to place highest priority on economic growth, particularly through assistance to the banking, housing, and construction industries, although they did provide unprecedented support for large-scale redistributive policies such as public works relief employment and public housing. Even in the area of policy that was redistributive, the administrative structure for the public housing program, for example, in the main worked against full program implementation by placing it in the hands of a local public authority separate from the city government and frequently controlled by antagonistic interests who tended to restrict it in various ways (Friedland, Piven, & Alford, 1984).

2. See Peterson, Rabe, and Wong (1986). These authors defined redistributive policies quite specifically as "those that benefit low-income or otherwise especially needy groups in the community" (p. 15). In contrast, they defined developmental policies more generally as "policies . . . intended to improve the economic position of a community in its competition with other areas" (p. 12). Their further discussion nevertheless reveals which interests are involved here as well, in observing that developmental policies have been most likely to succeed when key public figures such as Chicago's Mayor Daley have managed to coordinate "the interests of his political machine with those of downtown business leaders and the professional heads of public agencies" (p. 14). Typically, these powerful local elites have been among the major interests served by postwar urban developmental policies in restructuring cities (Levine, 1989; Stone & Sanders, 1987).

3. From mid-1934 to late 1935, FERA served as a combined cash relief and work relief program with states having the option of using FERA grants to fund either or both types of program, but this led to serious administrative problems. Following the termination of FERA, the Roosevelt administration made the decision to separate cash from work relief programs. Thereafter the funding and administration of cash relief programs were left to the states, while work relief continued as a federal responsibility (ACIR, 1982, p. 16).

4. On this see Gelfand (1975), who noted that "despite the traditional pleas for local autonomy, mayors considered the national character of WPA a vast improvement over FERA's reliance on the states" (p. 44).

5. As Hays (1985) noted, the successful 1935 court challenge by property owners in Louisville to federal use of eminent domain led to "a much more decentralized design for the public housing program, which helped to establish a more general pattern for federal programs—local planning and execution of projects utilizing federal dollars" (p. 175).

6. Outlining the objectives of PWA's housing program, PWA director Harold Ickes identified such familiar New Deal aims as reducing unemployment, eliminating slums, and providing decent, affordable housing to the poor. But the final objective he cited runs contrary to simplistic images of the New Deal as a highly centralized set of federal programs, namely, to encourage enactment of enabling legislation by the states "so as to make possible an early *decentralization* of the construction and operation of public-housing projects" (Bellush & Hausknecht, 1967, p. 7, emphasis added). As these authors observed on the same page, following this quote from an Ickes speech, "This is a paradigmatic statement not only of the New Deal's approach to the problems of an urban society but of every subsequent administration."

7. According to the Nader Congress Project (1975), FNMA presents an example of the indirect approach pursued by the federal government to reinforce the housing industry. Thus:

Rather than borrow money in the market at the lower interest rates available to it and then lend the proceeds out at below-market-interest rates to those in greatest need, the federal government insured private banks against loss on mortgage loans and then agreed to borrow in the market to pay off the private bankers whenever they wanted to get out of the mortgage business. (p. 156)

The Nader Congress Project (1975) authors were highly critical of this type of indirection, which they view as having contributed significantly to the creation of a fragmented, poorly coherent housing and urban policy system; as they put it, "Instead of opting for direct mechanisms to solve the nation's housing and urban problems, Congress persists in concocting weird, Rube Goldberg-type devices" (p. 173). In this regard, as Friedman (1967) observed, "Housing and urban development has become an empire of little gimmicks and bailiwicks, programs and subprograms, a bewildering, baffling congeries of devices, many of them motivated by the hope that the market can somehow be galvanized cheaply into life" (p. 368).

[5]

Restructuring American Urbanism

SUBURBANIZATION AND URBAN RENEWAL

This chapter focuses on the transitional period from the passage of the Housing Act of 1949 to the beginning of the War on Poverty in 1964. This period is marked by two trends of basic importance to postwar urban development. The first was the massive out-migration of generally middle-income population from older central cities, often following shifts in employment opportunities to nearby suburbs or to other regions. The second was the large-scale migration of minority and poor populations into these central cities, often into situations of great economic difficulty. Each of these major urban trends was influenced significantly by federal policies: *out-migration,* first by New Deal-created housing policies that encouraged suburbanization, then by wartime policies regarding the location of defense plants, and later by postwar policies in support of interstate highway development; *in-migration,* by New Deal policies that encouraged reduced farm production in order to reduce agricultural surpluses and World War II policies that contributed to the rapid mechanization of southern agriculture. The net result was the displacement of millions of farm laborers and tenants (Piven & Cloward, 1977) with neither local nor federal programs to assist in their retraining or relocation. Together these trends constitute what has been referred to as a *dual migration,* whose net effect was to weaken further the economic condition of older industrial cities already shaken by the Depression and the neglect of the war years that followed.

In response to numerous indications of urban economic and physical deterioration, particularly in older cities, the Housing Act of 1949 established a general framework for long-term change. As a housing act, it provided for clearance of central city slums and construction of public housing in their place. Yet at the same time it laid down the basis for a broad program of urban redevelopment that eventually would include extensive rebuilding of downtown commercially related areas as a key to the restructuring of central city economies. Thus the act can be seen as a two-pronged federal policy response aimed at dealing with mounting threats after World War II to both the social and economic viability of industrial era central cities. Those threats first became evident during the Depression, with growing awareness of the spread of urban slums and commercial blight associated with prolonged economic decline. Later, during the period of postwar decentralization, serious concerns regarding urban viability were heightened by the large-scale out-migration of city populations and economic resources associated with massive suburban development.

Consequently, the opening sections of this chapter briefly examine the process of large-scale suburbanization and the effects on it of several federal policies initiated since the New Deal. The major part of the chapter will focus on urban redevelopment as a strategy to rebuild the core of older central cities and its evolving redefinition from a program of slum clearance to one of *urban renewal*. In tracing this evolution, we will examine urban renewal not only as a program for coping with immediate problems of local economic viability but also as a long-term strategy for restructuring the older central city's economic, institutional, and residential core in the context of an emerging postindustrial national economy and of a changing metropolitan area economy.

SUBURBAN DEVELOPMENT AND URBAN DECLINE

Since at least the early 1920s, population and employment have undergone significant decentralization from older industrial northern cities. Following the end of World War II in 1945, the suburban environs of American central cities grew rapidly for several decades up until the 1970s. From 1950 to 1970, expansion of suburban population accounted for the bulk of metropolitan area population growth. Nationally, metropolitan areas grew by almost 45 million—from about 95 million to just under 140 million, with suburban rings absorbing over 75% of this growth for an increase of 35 million. As a result suburbia almost doubled its aggregate population, going from 41 million to 76 million, while central cities grew only from 54 million to 64 million, or less than 20% (Holleb, 1975, p. 12). Overall, the result was the emergence of the metropolitan area as the typical form of expanded, complex urbanization with contiguous areas linked together in networks of metropolitan development that extend across broad areas of the nation.

These trends indicate graphically the dynamic relationship between the two poles of the postwar metropolis: suburban development on the one side, attracting growing numbers of former city residents as well as industrial and business firms; and central city decline on the other. In metropolitan areas in general (but particularly in those associated with older industrial cities of the northeastern and north central regions), most growth of population, employment, new construction, and public improvements (schools, streets, lighting, water, and other utilities) took place in expanding suburban zones. In 1950 suburban areas accounted for a little over a quarter (27%) of the nation's population; in 1970 they contained 37%. During the same two decades the central city share of national population declined from 36% to 31%. In national perspective, by the 1970s suburbia had become the predominant living environment of Americans (Holleb, 1975, p. 12); by 1990, suburban population was approaching an absolute majority of national population.

What explains this historically unprecedented process of suburban growth? In the contemporary ecological perspective, several causal factors are cited in what has become by now a standard scenario of postwar suburban growth. These can be identified as push, pull, and facilitating factors. *Push factors* include changing conditions that affect central city population, such as the in-migration of poor and minority groups and the deterioration of urban physical stock and infrastructure, ranging from aging housing and industrial facilities to overloaded transit and school systems. *Pull factors* include residential preferences among urban middle-class families for life in greener, more spacious, more controllable surroundings than those of the large central city and lower costs of land, wages, and taxes for business firms.

The standard ecological scenario maintains that during the postwar period, push and pull factors reinforced one another to produce powerful incentives for out-migration from central cities. These were joined by technical *facilitating factors,* such as long-term improvements in highway transportation and long-distance communication, which contributed decisively to making possible this massive migration to suburbia.

If one were to assume that these factors were really the only important ones at work over the last four decades, it would be reasonable to argue as ecological theorists have that suburbanization was essentially a spontaneous free-market process, traceable to millions of private household and business decisions to relocate, based on push-pull factors that built up a vast reservoir of market demand. That reservoir of demand would be seen as having been released by technological advances that facilitated high-speed access to suburban residential and work sites on a large-scale basis.

However, such a scenario tends to neglect at least two other important variables. One is the role played by private corporate, financial, and real estate interests both in suburbanization and in urban renewal. The other is the role of governmental policy as a significant dimension of urban and metropolitan development, especially after World War II.

The role of corporate business and industrial organizations is evident in successive waves of suburbanization of industry dating from the turn of this century. Particularly during the postwar period, the out-migration of blue-collar employment in manufacturing industries such as autos and steel, traditional pillars of industrial city economies, has been a major factor contributing to the decline of older central cities. Large-scale manufacturing complexes established outside central cities often have served as important growth nodes in suburban development with residential areas growing up within commuting distance of these employment destinations. In addition, as postwar population expanded around suburban industrial nodes, a variety of commercial and service facilities began to "follow their customers to the suburbs" (Solomon, 1980, pp. 9-10). Thereafter, to the degree that those facilities clustered within growing shopping malls and office centers, they too began to act as nodes of further development. Over time, growing numbers of retail firms, service establishments, and professional offices moved out to the developing suburbs, thereby gaining access to a rapidly growing population with a wide variety of job skills and consumer tastes while avoiding the relatively high costs of central city location.

Overall, decisions by business firms (and particularly by larger firms) to locate beyond the city's boundaries become reflected in increased suburban employment opportunities for new residents. In turn, the expanding pool of suburban workers and consumers further encourages the migration of new business activity. At a certain point this sequence of events becomes self-propelling; but in its essentials, it can be argued that critical factors in the process include the locational decisions of industrial corporations with regard to location in suburban manufacturing complexes as well as the locational decisions of commercial investors investing in large-scale shopping malls and office facilities that will themselves tend to act as growth nodes.

At the level of the immediate locality, this process typically has been encouraged by a network of related interests sometimes referred to as the *local growth machine*, consisting of metropolitan area banks and mortgage lenders, real estate brokers, landowners, developers, lawyers, and the local construction industry, together with city government (Molotch, 1976). These interests, all of which stand to gain materially from local growth, constitute an interlinked division of labor in the acquisition, subdivision, and preparation of land and physical facilities for use by private business interests. The organizations in this network not only cooperate in the technical implementation of local growth but also join as a political-economic coalition in promoting or "boosting" the local place as a desirable site for business expansion and for residence by workforce and consumers.

By the 1960s the process of suburban growth had clearly become self-sustaining, and its impact on northern central cities was increasingly serious. However, the scenario of rapid suburbanization would not have been possible in the absence of another key factor often not given sufficient recognition in its own right. That factor has been the role of federal public policy in stimulating and augmenting the action of the other factors already mentioned.

SUBURBANIZATION AND FEDERAL POLICY

Suburbanization and Housing Policy

With the creation in 1934 of the Federal Housing Administration, the federal government undertook a major public policy that proved to have significant implications not only for the growth of suburbs but also for the future of central cities. In little more than a decade, FHA mortgage insurance was on the way to becoming an institutional key to home ownership, mainly in the suburbs, for millions of American families. In addition, at the end of World War II in 1945, FHA's insurance program served as a model for a similar program to help returning veterans purchase their own homes. Thus FHA's mortgage program was now joined by a mortgage program of the Veterans Administration (VA), making it possible for returning veterans to enjoy easy terms in buying a home with government again assuming the risk of default (National Commission on Urban Problems, 1968). Through these FHA and VA programs the federal government subsidized a massive wave of home building, most of it outside of central cities. Over time, such government programs profoundly affected the physical growth and the social structure of suburbia, as well as the economic and residential condition of central cities.

Having served the initial purpose of stimulating the home construction industry during the 1930s, the long-term function of federal mortgage insurance became primarily that of a program to support the construction of suburban housing. This was partly because both FHA and VA policies made it easier to obtain loans on new housing, as opposed to already existing housing. In addition, the FHA program was particularly oriented toward assisting the purchase of single-family dwellings, which require more land per dwelling unit than multiple-family housing. Over the postwar era, large lots of land were more readily available and lower priced in suburban areas than in central cities; in general, this systematically tended to favor new residential development in the suburbs (Shannon, 1983).

In spite of the agency's declared objective of expanding home ownership for American families, FHA-backed housing built in the suburbs was by no means intended for ownership by all groups. Until 1962, when President Kennedy signed an executive order to ban discrimination in federally aided housing, it was generally FHA policy to restrict loans to White borrowers; the agency went so far as to endorse as a model a racially restrictive housing covenant dating from the 1930s to protect all-White residential areas. For more than a generation of critically important community development, FHA actively discouraged investment in racially or ethnically mixed central city neighborhoods by advising its appraisers to give low eligibility ratings to properties located in areas that housed minority groups (Larson & Nikkel, 1979). It was not until the late 1960s that an FHA commissioner clearly declared that such redlining of whole areas was no longer permissible and that "FHA will not designate entire communities or areas as ineligible for participation in its mortgage insurance operations" (Fried, 1971, p. 70).

The other important socially discriminatory practice maintained during the crucial decades of postwar suburbanization until the early 1970s was that of income discrimination by FHA against government-backed loans to lower income groups. Eligibility requirements tended to exclude not only the unemployed poor but also lower income working-class families from qualifying for loan guarantees, even though in the aggregate these groups had the most serious housing needs in the nation. According to a report of the National Commission on Urban Problems (1968), groups in the lower 40% of the income scale received as of the late 1960s only a little more than 10% of FHA mortgage funds (p. 100). In fact, "up until the summer of 1967, FHA almost never insured mortgages on homes in slum districts, and did so very seldom in the gray areas which surrounded them" (Fried, 1971, p. 69).

The impact of this income bias, which overlapped with and reinforced the racial bias of FHA mortgage policy, can be judged by the magnitude of the program. Between FHA's establishment in the mid-1930s and the emergence of strong criticisms of its policies in the late 1960s, over 7 million housing units, or fully 20% of all privately financed nonfarm units in the nation, were constructed with FHA-backed mortgages (Fried, 1971, p. 68). The VA mortgage guarantee program, established in 1944 to provide GI home loans to returning World War II veterans, proved somewhat more accessible to lower middle-class groups. Almost 30% of VA mortgages in 1966 went to families having incomes between $4,800 and $6,000, as against only 10% of FHA mortgages that went to such families in the preceding year. Nonetheless, both the FHA and VA mortgage programs tended to exclude the poor and the near poor; in each case, as of 1966 less than 2% of mortgages went to families with incomes of $3600 or lower (p. 85). Overall, it may be said that both programs contributed to a policy of selective suburbanization, favoring White middle-class movement to the suburbs. As Solomon (1980) noted: "The middle-class demand for low-density living was reinforced by government-sponsored FHA and VA insured mortgages and favorable tax policies which reduced the relative price of the single-family, owner occupied units, largely concentrated in the suburbs" (p. 12).

By contrast, there was little federal subsidy for multiple-unit structures typical of apartment dwelling in central cities. The net result was a strong bias in favor of middle-class ownership of new homes in the suburbs. At the same time, within central cities New Deal-initiated public housing projects became an important concentration point for poor and minority populations, adding to the push factors contributing to rapid suburbanization identified earlier in this chapter.

By the early 1960s federal housing programs had increased significantly the volume of mortgage credit, bringing moderate-income families with relatively little savings into the housing market on a large scale. From the point of view of young families seeking affordable new housing, FHA's program was a considerable success. Yet considered in relation to the situation of postwar American central cities, the program subsidized "a massive middle-class migration to the suburbs." Federal housing policy, consisting essentially of the FHA program and the public housing program, "contributed directly to . . . the nation's most serious urban dilemma: isolation of the poor in central city ghettos shut off from job or educational opportunities" that were now increasingly to be found in the suburbs (Nader Congress Project, 1975, p. 161).

Suburbanization and Defense Policy

Postwar suburban growth was assisted significantly also by a federal program to construct war production plants during World War II, initiated by the creation of the War Production Board (WPB) in January 1942 shortly after America's entry into the war. The ensuing decisions of this board "probably did more than any subsequent federal program to reshape urban America" (Kantor, 1988, p. 199). Through the WPB, the federal government expanded the nation's industrial capacity by nearly 60%, with the board spending over $23 billion in the space of a few years to supplement existing manufacturing facilities whose total value had been placed at $39.5 billion in 1939 just prior to the war (Mollenkopf, 1983, p. 104). It subsequently was observed that this enormous investment paid for "the greatest increment to manufacturing capital recorded in modern industrial history" (Civilian Production Administration, 1946, p. 40).

The federal expenditures that were involved completely overshadowed the public works programs of the New Deal; the PWA, which was the developmental program specifically devoted to producing large-scale public infrastructure, spent only $4.8 billion, or about one fifth of the cost of WPB's capital investments. But the significance of WPB's investments for American urban development was not simply that they dwarfed those of the New Deal. What made them critical to future urban development was the pattern of locational decisions made by the WPB (Kantor, 1988, p. 199). Not an agency of the New Deal, WPB was run basically by corporate executives on leave from the very largest industrial firms; the relatively few representatives of organized labor acted only in an advisory role.

Under the control of a business-oriented WPB, the massive program of wartime reindustrialization tended to place new industrial facilities in suburban and Sun Belt locales, beyond the established orbit of largely northern urban-centered industrial unions. The effect was to sponsor the movement of urban working-class constituencies out of the older industrial cities into their surrounding suburban areas or into other regions in southern and southwestern states, typified by more conservative governments and by less receptive attitudes toward union organizations. Overall this resulted in reducing the numbers and influence of the organized urban liberal voting bloc, with important long-term implications for the making of urban policy.

Suburbanization and Highway Policy

Another federal program with significant consequences for suburbanization was created by the Republican administration of President Dwight D. Eisenhower through the Interstate Highway Act of 1956. This program had tremendous influence on the overall pattern of postwar urban and suburban development. It not only promoted the growth of new metropolitan areas nationwide; it also stimulated rapid suburbanization as an important conservative response to the focusing of federal policy on the future development of central cities (particularly older, larger, more deteriorated industrial cities with their heavily organized labor and Democratic constituencies) that had been inscribed into the Housing Act of 1949.

The building of the interstate network can be seen now as a critical element in accelerating the trends toward suburbanization that had been apparent around major urban centers as early as the 1920s. The decision to proceed with this massive undertaking, the largest public works project in American history, was promoted vigorously during the early 1950s by the Eisenhower administration with the support of the auto, oil, and related auxiliary industries (rubber, glass, steel, and cement), which constituted a powerful postwar highway lobby, and was adopted by Congress in the Interstate Highway Act of 1956.

This act established a Highway Trust Fund to be financed from a federal tax on gasoline. From this fund, whose growth was premised on the expansion of automotive transportation, the federal government would pay 90% of local highway construction costs, thereby providing a powerful incentive for program participation by states and localities. The network was planned to include over 40,000 miles of interstate and local highways, of which more than 5,000 miles were allocated to metropolitan areas, shaping their development for decades to come (Muller, 1981, p. 52). With the construction of the interstate network, middle-class central city residents looking for newer housing or a better neighborhood were no longer confined to what the city had to offer; they now had ready access to expanding new suburban developments. At the same time, a variety of business firms, manufacturing plants, and real estate developers all found open to them the broad expanses of suburbia. No longer tied to locations near existing streetcar routes or railroad lines, they were freed from the pattern of urban-centered location associated with prewar urban-industrial development.

Only the federal government had the authority and resources to take a step of such magnitude. With the Interstate Highway Act of 1956, the governmental context for urban development was no longer the transportation policies of particular states or localities; rather, it became a national context, ultimately affecting every important urban center in the country.

In brief, the federal highway program opened the way to a new form of urbanism—the low-density, multicentered metropolis—with new spatial and social relationships, including new distributions of population and residential settlements. Highway development affected not only the areas around central cities but those cities themselves. Highway construction through the countryside helped expand urban development to metropolitan scale through encouraging suburban growth around central cities. At the same time, expressway development was brought also into the central city, providing the growing suburban population with commuting access to central city employment, shopping, and recreation.

The routing of the new express highways directly through urban areas had important impacts on central cities. Not least was the result that many minority and working-class neighborhoods were cut through and destroyed as viable communities in order to make way for the new multiple-lane, limited access roadbeds. This in turn meant the out-migration of former neighborhood residents into nearby areas, converting once stable low- and moderate-income communities into deteriorating slums.

As of the late 1950s and 1960s, many moderate-income neighborhoods lying between older residential areas of the inner city and the newer middle-class areas of the outer city and suburbs came to be known as *gray areas*. Contributing to the pressure on such areas was the spillover of migrants from highway development, added to that from downtown urban renewal. The severe local impacts of both these programs swelled the flow of migration between neighborhoods, sending ripples throughout the physical and social structure of inner-city neighborhoods and the gray areas of the next zone and causing serious stress in the more socially or economically fragile of these neighborhoods.

On balance, the major impact of urban expressway construction has been "to encourage—in effect to subsidize—further movement of industry, commerce, and relatively well-off residents (mostly white) from the inner city" (Banfield, 1970, pp. 14-15). The effect on the inner city has been to reduce accessibility to the many industrial and service jobs that have located in the suburbs, as well as to further segregate inner-city communities from the rest of the city, its institutions, and facilities. In addition, highway construction and related urban expressway development have competed for riders and funding against urban public transit systems on which lower income city residents are highly dependent. Overall, federal policy for developing a national highway system with expressway links to metropolitan centers clearly contributed to the decline of central cities at the same time it supported the rapid development of the suburbs.

URBAN RENEWAL AND
THE RESTRUCTURING OF AMERICAN CITIES

As of 1920 American cities had arrived at a historical watershed with the census revealing that a majority of national population lived in urban places. Big cities such as New York and Chicago displayed new high levels of population and employment, beginning a decade of frenzied prosperity that ended in the crash of 1929. This was followed by almost two decades of urban neglect during the national crises of the Great Depression and global war. As a consequence, by the late 1940s older industrial cities in particular, generally concentrated in the Northeast and Midwest, were threatened by physical decay and economic decline. A rising volume of opinion then argued that such cities needed something special to help resuscitate them, to make them vital and competitive again, especially in face of the challenge posed by rapid suburban expansion.

However, long before the end of World War II and the beginning of postwar suburbanization the problems of the built physical environment of industrial era cities had begun to accumulate. Private investment in housing declined abruptly due to the Depression of the 1930s and remained low owing to the diversion of resources to the war effort during the 1940s. As a consequence, deterioration had serious effects, particularly on the older housing stock of central cities. The most severe impacts were experienced in inner-city districts in which existing slums became dilapidated beyond repair and deterioration spread to adjacent residential areas, converting them to new slums. This led to a growing shortage of safe and sanitary housing, especially in industrial cities crowded with defense workers.

With the return home of millions of veterans, many of whom married and formed households with a rapidly increasing number of children (later to be known as the baby boom generation), the housing situation assumed critical proportions. In addition, although some notable public improvements were made through the public works programs of the New Deal during this period, those programs were terminated soon after the nation's entry into World War II. As a result, many central business districts took on a shabby and blighted aspect, particularly as commercial investment also was withheld. By the end of the war, older central cities thus faced a complex set of problems stemming from almost two decades of neglect: a serious housing shortage, the intensification and spread of urban slums, and deteriorating central business districts.

By the late 1940s these problems had come to be perceived as sufficiently critical to influence the urban policy agenda at both the local and national level. The return of large numbers of veterans to urban centers made it impossible to ignore any longer the unsanitary, substandard, overcrowded housing conditions particularly characteristic of lower income areas of older cities, and these now came to the forefront as a public issue. There was strong sentiment, particularly among big-city publics, in favor of the need for housing reform (Mollenkopf, 1983, p. 75). A broad coalition including veterans' organizations, labor unions, community organizations, and religious groups demanded that government take action in support of provision of decent affordable housing. The rise of these popular demands soon after World War II meant that housing problems received high priority among the issues addressed by Democratic incumbent Truman in his 1948 campaign to retain the presidency. At the same time, during this period there was growing support in the urban business community for physical redevelopment, but of a type quite different from that advocated by housing reformers. In this way, significant urban issues that had been put aside for several decades now were due to be brought to the public policy system for some type of resolution.

Through the Housing Act of 1949 the urban policy system attempted to resolve these different perspectives on the future of the postwar American city. Before discussing the act and how it was modified during the following two decades, let us consider briefly the dimensions of the social-spatial context in which the major policy choices were made with long-term consequences for the development of the postwar central city.

Dimensions of Central City Restructuring

Earlier we have examined postwar suburbanization with regard to some of its basic impacts on central cities. Suburbanization can be viewed as the exterior dimension of the postwar transformation of American cities, involving the spatial decentralization of urban population and of certain established economic functions, especially those associated with large-scale industry and commerce. Much of this was visible to interested observers as it took shape physically in the form of suburban residential subdivisions, industrial parks, and shopping malls. Indeed, it was a major subject of observation and comment by contemporary ecologists working in the postwar period.

However, a less obvious side to postwar suburbanization was the growing economic abandonment of older cities as private enterprise—such as lenders, builders, and real estate firms—was attracted to business and investment opportunities in the suburbs. In general, rising private investment in suburbs meant relative disinvestment in central cities (Bradford & Rubinowitz, 1975; Burchell et al., 1984), although this was not usually evident until its effects were manifested physically in residential deterioration and commercial decline in urban neighborhoods and business areas. Also not readily evident, yet still quite real in its effects, was the circumstance that such urban disinvestment was reinforced frequently by federal policies: by mortgage policies that opened opportunities for investment in suburban home ownership mainly for the broad urban middle class and by federal defense and highway construction policies that facilitated expanded business and real estate investment in suburban development.

The interior dimension of the postwar urban transformation concerns the development of new core activities within city economies and the corresponding physical restructuring of older cities. Over the long term such cities hoped to use urban redevelopment to restructure their economies, encourage new investment, and strengthen their fiscal positions. After World War II older large industrial cities entered an accelerating process of transition from centers of industrial production to centers for the production of specialized business, professional, and social services. As large-scale industry shifted to the suburbs or to other regions in search of lower priced land and labor force, older central cities tended to become hosts to expansion of corporate office activities and advanced business services. Accordingly, at the same time that industrial employment and employment in consumer services were declining, there was rapid expansion in specialized business and government services. A study of data for 11 large cities from 1950 to 1967 showed that although these cities lost 400,000 jobs in manufacturing and commerce, they added 1 million jobs in business, financial, professional, and public services (Gantz, 1972, p. 74). Another study of 12 large northern cities from 1948 to 1977 showed severe job losses in manufacturing, wholesale, and retail sectors amounting to a total of over 2.4 million jobs; over the same period, however, selected service jobs in these cities grew by over 300,000 (Kasarda, 1985, pp. 43-45).

In short, older big cities have undergone a major transition from producing goods to providing services, particularly administrative and professional services prompted by the growth in scale of modern business and governmental organizations. The effects of this transition perhaps can be summed up in the observation that the day of the factory-centered city is over. Increasingly the restructured American city is a postindustrial city with an administrative complex of corporate and government office facilities at its core, accompanied by banks, trade centers, hotels, restaurants, department stores, and related business and personal service establishments, with a marked decline in its concentration of industrial firms and employment.[1]

Organizational Restructuring

During the postwar period many large corporations in search of lower costs shifted the locus of their production facilities out to the suburbs or to other regions such as the South or out of the country. At the same time, corporate control mechanisms became more organizationally concentrated and physically centralized than ever before, as evidenced by the expansion of corporate office facilities in major central cities. The bigger the firm, the more this tended to be the case. With the growth in scale of the firm, its administrative, informational, and planning requirements tended to grow; this was particularly true for the higher decision-making levels of the firm (Friedland, 1982).

Pressure on corporate elites to cope with the rapidly changing postwar economic environment and the need for improved coordination within extended corporate networks have led to the development of complex interurban and urban-suburban organizational linkages between the component units of multilocational corporations. Generally these linkages involve administrative services or the transmission of control decisions and coordinating information from central corporate headquarters to subordinate units (Pred, 1977). Organizational growth thus has prompted the growth of the managerial components of large corporate firms. It has encouraged also the emergence of organizations outside these firms dealing in highly specialized business services that contribute to corporate administrative capacities.

These developments are related to the postwar emergence of several related institutional clusters characteristic of the political economy of the postindustrial city. First, within the *business sector* these institutions include not only large corporate and financial firms but also a variety of advanced business service firms in finance, law, accounting, advertising, and other specialties that provide essential support to the work of large businesses. Together these constitute the dynamic core of the contemporary urban business sector. Second, within the *public sector* the relevant institutions encompass local government agencies and public authorities involved in providing business-related infrastructure (such as urban renewal programs) or social services (such as job training programs); they also include federal agencies that provide financial and technical resources to local counterpart agencies. In combination, these organizations structure the operation of the urban intergovernmental program delivery system in the postindustrial city (Friedland, 1982; Mollenkopf, 1983).

In addition, in large central cities a distinctive segment of the private sector is now composed of private nonprofit organizations such as hospitals, universities, and private social agencies, all of which have grown significantly during the postwar era. This can be viewed as the *intermediate sector* of the contemporary urban economy because in several respects it stands between those enterprises of the private sector that are for-profit and the nonprofit agencies that make up the public sector (Mollenkopf, 1983). On one hand, it plays a complementary role to the public sector by providing services on a contract basis directly to public agency clients and by otherwise filling gaps in locally needed health or social services; on the other, its university component complements the business sector by providing to it advanced scientific/technical training, research, and related consultant services, often of an innovative nature.

Physical Restructuring

Postwar growth of the new urban business complexes often has involved the expansion of corporate office space in central city CBDs conveniently close to auxiliary business services. Locating in the CBD allowed firms to take advantage of the highly specialized banking, accounting, advertising, legal, and other business services available there that might often be too costly to produce within the firm. CBD location also allowed for extensive face-to-face interaction between the firm's executives and other business executives, as well as between corporate technical staffs and providers of advanced business services—whether at the office conference table or the private downtown luncheon. In these respects, the rise of the new central office complexes represents in good part the physical expression of the importance of reliable current information and analysis in today's large business firms as embodied in the accessing, processing, and transmission of information for purposes ranging from routine tracking of everyday business data to high-level planning and policy making. The unique significance of the central city for these organizational functions is that it "continues to offer a location where channels carrying information of potential relevance to corporate offices are concentrated. This information is produced, stored, and exchanged through a dense interorganizational network. Corporate planning depends on its linkages into this network through service contracts, interlocking directorates, and social contacts" (Friedland, 1982, p. 65).

It is conceivable that CBD renewal to satisfy the physical requirements of corporate operation could have been achieved through private purchase of land sites. However, in prime downtown locations and nearby areas such sites typically were expensive and difficult to assemble in parcels large enough to accommodate well-designed projects. As a result, the construction of major private office facilities increasingly was accomplished in conjunction with governmental programs that paid for site acquisition and clearance as well as public infrastructure ranging from water and sewage utilities to cobblestone sidewalks and stylish streetlights.

Similarly, the emergence of new government office complexes in central cities since World War II has produced considerable downtown land clearance, particularly in large cities. This reflects the tendency following the New Deal of older central cities to provide a broad range of educational, health, and social services that go beyond the traditional basic services such as police, fire, and sanitation typical of cities early in this century. The resulting growth in public employment has been matched by the rise of government office complexes to accommodate the expanding public agencies and programs. Thus the consequence for the city as a built environment has been the simultaneous physical expansion of public and private organizational complexes entailing major postwar redevelopment activities most frequently in and around urban central business and administrative districts.

URBAN RENEWAL:
FROM SLUM CLEARANCE TO
DOWNTOWN REDEVELOPMENT

The development of a postindustrial national political economy has been characterized by the rise of large-scale business and governmental institutions and of the physical facilities—typically located in large central cities—to serve them. The administrative informational requirements of the elites heading these institutions include the capacity for rapid acquisition and processing of large quantities of organizationally relevant data. They also require ready personal access to primary information and decision networks, as well as organizational access to a wide variety of administrative support services—all which have been uniquely available in central cities. Such are some of the organizational factors that have been influential in shaping the central office locational decisions of major private and public institutions during much of the postwar period. They have acted also as considerations in favor of the use of public resources to support redevelopment of the central business district as a modernized high-rise administrative office and professional facilities center, a key aspect of postwar urban renewal.

Evolution of Urban Renewal

Three phases can be identified in the evolution of the federal urban renewal program. The first spanned the 5-year period following the passage of the Housing Act of 1949. This early period was a tentative one in which the specific character of urban renewal (commonly referred to at that time by the generic term *urban redevelopment*) was in question and did not begin to take on a definitive form until passage of the Housing Act of 1954 (Gelfand, 1975). During this period, to the extent that urban renewal had an identifiable goal, it was that of clearing away residential slum areas near central business districts presumably to make land available mainly for construction of affordable housing units for low- and moderate-income households. However, the program as structured during this period failed to attract many private developers; it cleared only a small amount of land and did not result in much redevelopment. I refer to this simply as the *slum clearance phase* of renewal.

The second period covered the next decade and a half from the Housing Act of 1954 to that of 1968. Due to amendments incorporated in the 1954 act and in later legislation, by 1960 the program was attracting the growing participation of private developers. Even before that it had assumed its conventionally defined character as urban renewal: a program mainly responsive to the interrelated elite interests of private commercial entrepreneurs, local officials acting as public renewal entrepreneurs, and representatives of major central city institutions. Emphasis on clearance of renewal sites (rather than rehabilitation, for example) tended to involve large-scale demolition of existing properties, especially in low-income residential neighborhoods bordering on CBDs. During the 1960s redevelopment on cleared sites was increasingly commercial and institutional in nature, particularly taking the form of office building construction or expansion of core professional institutions such as hospital and university complexes. Overall, most residential redevelopment during this phase produced either middle-income or luxury housing units. This can be understood as the *entrepreneurial phase* of urban renewal.

The third period covers the next 6 years from the Housing Act of 1968 to the abolition of urban renewal as a free-standing program in 1974. I refer to this as the *Creative Federalism phase* of renewal, reflecting the introduction of housing and community development objectives established under the War on Poverty, including a shift in emphasis to rehabilitation in residential areas. Following this phase, the urban renewal program was ended through absorption into a new type of community development program no longer targeted to older industrial cities or their poverty areas, established in 1974 under the Nixon administration's New Federalism (described in Chapter 7).

From Slum Clearance to Urban Renewal

The shift in emphasis in the urban renewal program from one that seemed focused initially on replacing inner-city slums by low-income housing to one increasingly dedicated to nonresidential development occurred through a process of amendments to the 1949 Housing Act over the next decade and a half. During the years immediately following the act, there was little interest shown by either real estate developers or corporate interests in what appeared to be primarily a program to develop low-rent housing near downtown commercial areas. In addition, potential developers perceived the risks of investing in renewal areas as being very high, particularly because the Federal Housing Administration, faithful to its conservative practice of favoring new housing construction elsewhere than older low-income urban areas, refused to issue mortgage insurance in such neighborhoods. Thus something had to be done to assure investor confidence regarding investment in slum areas.

Incentives to private investors came in the form of a series of amendments to the 1949 act that allowed communities to shift a growing percentage of project funds from the construction of housing to commercial redevelopment. The 1954 amendment established the precedent by allowing the use of 10% of project funds for nonresidential development; in 1959, this was raised to 30%, and in 1965, to 35% (Friedland, 1982, p. 80). Actually, these nominal percentages understate the extent to which grant funds could be diverted from the initial housing focus of the act. In practice, over half of federal funding could be diverted from housing to other purposes.

Any renewal project that allocated 51 percent or more of its funds to housing was designated by the federal administrators as a "100 percent housing" project. By manipulating this definition, local authorities could allocate as much as two-thirds of their funds for commercial development, but still remain within federal guidelines. (Judd, 1988, p. 269)

Such leniency in the standards for implementing what had promised to be a program for improving the living conditions of the inner-city poor followed from a local renewal perspective dominated by political and economic elites. Whereas local politicians were reluctant to become identified with a position of advocacy for public housing, business leaders were outspoken in asserting that the highest priority in dealing with urban deterioration was the revitalization of the downtown business district. In time the notion that downtown business renewal was the best way to attack the root causes of urban decline became accepted widely by political officials and community influentials as well. In the absence of an organized, politically articulate constituency for publicly subsidized housing, the provision of decent housing to the poor was downgraded seriously among the several objectives of urban renewal.

In part, the abandonment of the goal of affordable housing for the urban poor was the product of campaigns against public housing conducted with the active encouragement of the National Association of Real Estate Boards through its local chapters in dozens of cities during the early 1950s. Following the passage of the Housing Act in 1949, local referenda resulted in the rejection of public housing by 40 communities, including such large cities as Houston and Los Angeles (Hartman, 1975). Such referenda helped to keep public housing out of the newer big cities of the Southwest and West and maintained its concentration in the larger industrial cities of the Northeast and the north central region, such as New York, Philadelphia, Baltimore, Chicago, and Detroit.

At the federal level, two consecutive developments confirmed the downgrading of public housing as an objective of the 1949 act. First, by the mid-1950s congressional funding authorizations for public housing were cut back sharply based on the 1954 amendment of the act. Subsequently, during the 1960s Congress established a number of new federal programs that subsidized low-income housing but were legally and administratively separate from the urban renewal program (Friedland, 1982, pp. 80-81). (These included subsidies to builders in the form of low-interest loans for low-rent housing construction voted by Congress in 1961 and subsidies to tenants in the form of rent supplements by legislation in 1965.) Such administrative segregation of low-income housing programs reflected the fact that these programs had lost the high priority associated with the renewal program as the major means of central city restructuring.

By the mid-1950s local campaigns against public housing and federal reductions in program authorizations had begun to make it clear to municipal officials that strong support for public housing was politically a losing proposition. President Eisenhower drove the point home by appointing "a former Republican congressman who had voted against the 1949 Housing Act because of its public housing provisions" as the head of the Housing and Home Finance Administration (Mollenkopf, 1983, p. 117); he also appointed a member of the National Association of Real Estate Boards to head the federal public housing program.

At the same time, an urban political agenda that emphasized economic redevelopment held out the hope of improving both business profits and personal income for residents, as well as expanding public resources available to fund a wide spectrum of local public programs in areas such as education, housing, and public works. In business-oriented urban renewal there was also the promise of a significant political boost to the position of the enterprising activist mayor who associated himself or herself with improving the financial status and public image of an economically and physically revitalized urban core. Finally, control of the urban renewal bureaucracy—with its power to strategically affect key developmental processes and the city's revenue base—could give new leverage to an alert entrepreneurial mayor in confrontation with other bureaucratic fiefdoms whose own budgets and programs stood to benefit from expanded city revenue.

Progrowth Coalitions and the Renewal Establishment

The 1954 amendment signaled what was to be a continuing shift in the priorities of the 1949 Housing Act. With the deemphasis of housing—particularly low-income housing—as a priority and new readiness to use federal funds for nonresidential redevelopment, urban renewal gained momentum as local developers came forward and local politicians showed new interest in the program (Friedland, 1982). By giving this new type of renewal a central place in the administration's agenda, an enterprising mayor could make important strides toward securing his or her political future. A leading example is Mayor Richard Lee in New Haven, Connecticut, who during the mid-1950s fashioned a broad political alliance around the promise of urban renewal. This alliance was composed of downtown business interests, construction unions, the city's planning and renewal agencies, and the local Democratic party as well as representatives of Yale University, and it had strong connections with federal urban renewal agencies.

The major axis of the alliance was the coalition between local public officials, led by Mayor Lee, and local businesspeople with strong interests in

CBD redevelopment. With Lee, the city government's chief executive, acting as the political-administrative link between public renewal planners and private business interests, the basis for an *urban progrowth coalition* was established.[2] In essence, this coalition was an executive-centered public-private partnership (Salisbury, 1964) purposely established to foster a renewal program oriented to central business interests, to the growth demands of the city's leading professional institutions, and to the requirements of public officials seeking to increase the city's tax base in order to produce sufficient revenue to deal with growing urban social problems.

The progrowth coalition, combining governmental and business elites and led by a strong chief city executive, represented a new layer of the local political economy whose formation was encouraged and supported by federal urban renewal program activity. The governmental organizations with which it was related specially and whose policies it significantly influenced constituted a powerful renewal establishment (Fainstein et al., 1983). This establishment, generally brought together through the efforts of the city executive, often consisted of a redevelopment agency and several auxiliary agencies. The latter might include such preexisting agencies as city planning, public works, and transportation (each of which was likely to receive infusions of renewal funding) and perhaps a newly established department of development coordination. Although this complex of public agencies did not displace the routine bureaucratic government of the city, they were positioned to ride atop the regular bureaucracy as a type of upper organizational tier with special powers for planning and implementing core city renewal.

Urban renewal establishments created in the middle and late 1950s often grew on the basis of resources funneled into postwar city government by way of the vertical connection created by federal urban development programs. Nourished by federal grants, the urban renewal establishment manifested clearly the joining together of major interests and institutions of the bureaucratic city state produced by urban reform and the urban intergovernmental policy system produced by the New Deal. Until the mid-1960s powerful renewal bureaucracies generally operated as they deemed fit without any effective challenge to the policy of large-scale "bulldozer clearance," which led to widespread resident displacement and neighborhood destruction. Only amid the popular stirrings associated with the social turbulence of the later 1960s did renewal establishments meet with organized neighborhood resistance sufficient to produce significant modifications in the renewal program.

Overall, the creation of local renewal establishments exemplifies the patterns of vertical-horizontal relationship that came to characterize urban policy during the entrepreneurial phase of renewal. Such renewal establishments can be understood as the product of interaction between the evolving federal renewal program (vertical, intergovernmental axis) and local progrowth elite coalitions (horizontal, local axis). The federal Housing Act of 1949 first provided a broad framework within which city governments could begin laying the organizational bases for local implementation of renewal. With encouragement by the amendments of 1954 and after, full-fledged renewal establishments were created and took on the distinctive operating pattern of special responsiveness to elite local constituencies such as downtown business and institutional interests as well as to the federal renewal bureaucracy that provided funds and technical assistance for renewal operations. Let us turn to examination of these patterns of operation and of the local renewal program structures associated with them.

URBAN RENEWAL:
PROGRAM OPERATION AND ORGANIZATION

As noted, the 1949 Housing Act and amendments established the general framework for federal involvement in urban redevelopment during most of the next two decades until the late 1960s. Under those provisions the redevelopment program was federally funded but locally administered. Title I of the act empowered the Housing and Home Finance Agency to assist local efforts in removing slums and blighted areas. Through this agency the federal government offered grants to help city government meet the cost of public subsidies in support of the renewal process.

The basic procedures of the redevelopment program can be summarized as follows. First, city government either designates itself as the local renewal authority or appoints a special agency (Lindbloom & Farrah, 1968). That agency has the authority to select renewal sites after identifying them as being "blighted" (an ambiguous term implying an unhealthy appearance of disrepair and unsightliness, which allows the taking of land and property not necessarily so deteriorated as actually to be in slum condition). Using the power of eminent domain it can purchase a given site at a "fair market value" whether or not the owners wish to sell at that price. It then contracts to have the site physically cleared of existing structures and prepared for sale to a private developer, usually at a price considerably lower than the developer would have to pay directly in the private market (Greer, 1962, p. 84).

In short, the costs incurred by the local renewal agency for purchasing and clearing the project site typically are greater than what it receives on sale of the site for redevelopment. This difference—known as "the renewal write-down"—constitutes a public subsidy to the developer and usually would be too costly to be carried by local government alone. However, under Title I of the Act, the federal government pays two thirds of the "write-down" thereby making the renewal process possible (Friedland, 1982, p. 80). The remaining one third of the write-down is paid by the locality, generally in the form of public facilities provided at the renewal site. Thus, operating within the framework of a decentralized program structure, the local renewal agency uses its powers of eminent domain to assemble a suitable site (typically by combining smaller separately owned parcels, which often would be difficult to accomplish if left to a private real estate agent negotiating with multiple private owners), and the federal and local government share the costs of public subsidy for site acquisition and clearance.

Although the urban renewal program involved a good deal more than just housing, its operation can be understood better by comparison with the two major housing programs that preceded it—the FHA mortgage program and the public housing program. FHA mortgage insurance was a federally subsidized program administered by local private financial institutions; public housing was a federally subsidized program administered by local public authorities. Under each of these programs implementation thus essentially was decentralized. For the most part the renewal program's structure was decentralized also in that the initiative for program involvement and decisions about specific project actions came from the local level while Washington acted as the federal bank providing program funding.

However, in pursuit of its developmental goals the urban renewal program incorporated a more complex pattern of interorganizational relations than did either the FHA insurance program or the public housing program. Its program relationships extended not just between the federal government and private industry (as in the case of FHA) or just between Washington and the local public agencies (as with public housing) but between each of them simultaneously. Considered in these terms, urban renewal had a dual aspect: It was an intergovernmentally decentralized program federally subsidized to support redevelopment; and it was also an intersectoral program connecting public and private sectors in a government-business partnership for redevelopment. Along the lines of the basic model first established by the Public Works Administration, the federal renewal program delegated the implementation of policy objectives to its partners at the local level—first to the members of the local renewal agency and through them to the private sector developers who participated in the program. As a consequence, local operation of the urban renewal program depended on close cooperation with private developers but largely was independent of city government and of the local public housing agency.

At the local level, the renewal agency typically was committed to activities that would show visible and relatively measurable results both to the local community and to the federal renewal agency. Despite the early promise of the program to pursue a predominantly residential approach, public improvement of slum areas by providing better housing or community amenities for residents had little support among influentials in the community decision system. Other goals tended to attract far more attention among local business and political leaders. Foremost among these was the physical redevelopment of deteriorated business districts and nearby blighted residential areas, providing land for corporate office towers, hotels, theaters, convention centers, and the like, to provide the city with added tax revenues. On this there was a considerable degree of elite consensus. To the extent that residential renewal was embraced by local influentials, it was on the grounds that it would be used to attract middle-class population back to inner-city areas. Such population could be expected to provide white-collar workers for the corporate towers and shoppers and patrons for the hotels, theaters, and other amenities that would appeal to upscale residents. This again would contribute to increasing the tax base.

In focusing on these economic goals, the renewal program produced some serious social costs impacting very severely on existing residents in clearance areas, particularly in low-income non-White neighborhoods adjacent to the CBD. By the 1960s this began to arouse local opposition, as reflected in charges that urban renewal was in fact "a program which under the guise of slum clearance is really a program of Negro clearance" (Wilson, 1967, p. 288). From the perspective of supporters of business-oriented renewal, the operational issue became finding a means to protect urban renewal from popular opposition. One avenue was to persuade big-city mayors to embrace the program as a federally subsidized means to reviving their city's economic position. Acceptance by activist, often Democratic, big-city mayors who already had political ties to minority communities tended to give the program a certain public legitimacy that it otherwise would have lacked. At the same time, however, such an openly political approach to protecting the program could leave mayors vulnerable to public demonstrations or electoral action by minority and low-income neighborhood constituencies; this was a possibility whose likelihood would increase with the growing local organization and political mobilization of such communities (Friedland, 1982).

Program Administration and the Bias of Renewal Organization

Another approach that can be thought of as the structural or organizational approach and that tended to be more effectively resistant to public protest was to organize the urban renewal program at the local level in such manner as to shield it from direct political challenge. In many large cities a central aspect of the organizational approach to shielding or insulating the renewal program from public challenge was the administrative separation of the public housing agency from the local renewal agency. This meant that low-income groups adversely affected by renewal activities would have to pursue two separate channels if they wanted to do more than merely protest displacement from renewal sites. In addition to picketing or otherwise lodging their protest with the renewal agency, they would have to deal also with the public housing agency in order to lobby for low-cost replacement housing on site or elsewhere in the city. However, besides having little influence over renewal agency policies, local public housing agencies often had to rely for usable housing sites on the renewal authority, which was a key agency in regard to the clearance of land on which any type of redevelopment might take place in the city.

The widespread practice of separating the administration of public housing from urban renewal was no accident. It flowed from the viewpoint that urban renewal should be a program whose first priority was reconstruction of the central business district, rather than rehousing the inner-city poor. This view was advanced particularly by the Urban Land Institute (ULI). Established as a research arm of the National Association of Real Estate Boards in 1939, ULI became a major national lobbying organization for large property owners, real estate agents, and developers with interests in central city redevelopment. ULI spokespersons endorsed administrative separation at congressional hearings on the 1949 Housing Act; the organization also lobbied at the state level to incorporate this principle into state enabling legislation for local renewal. The U.S. Chamber of Commerce held similar views and worked in tandem with ULI to promulgate its positions. The chamber brought with it business support of ULI's attempts to correct the impression, widely held during the early 1950s, that urban redevelopment would be largely a housing program with a significant role for public housing. Following campaigns at the state and national levels, administrative separation was attained federally through the Housing Act of 1954, which created an Urban Renewal Administration separate from other housing-related programs in HHFA (Mollenkopf, 1983).

Separation from the public housing agency gave considerable autonomy to the local renewal authority in relation to the one local agency that might have fostered an alternative approach to business-oriented central city redevelopment. In addition, the renewal authority's direct relationship with the federal government gave it technical and informational advantages in interpreting complex federal regulations governing the renewal process. This provided strategic leverage, particularly in negotiations with city officials over specific redevelopment decisions (Friedland, 1982).

Another source of autonomy was the superior administrative resources at the disposal of local urban renewal agencies; federally funded renewal agencies tended to be staffed fully enough to be able to perform most of the planning, execution, and evaluation of their own projects. This allowed them to avoid significant reliance on other local agencies such as the planning commission, which might not necessarily agree with their plans. Moreover, the federal regional administrators also "lacked staff capacity to do either the long-range planning or adequately monitor" the growing volume of renewal projects (Friedland, 1982, p. 99). As a result, local redevelopment agencies largely were able to avoid detailed review or supervision by either federal or local agencies.

Although local renewal agencies were relatively autonomous from other public agencies, they were heavily dependent on private developers for implementation of renewal; as public agencies they were themselves excluded by the 1949 act from construction for the private market. The renewal agency therefore had to pay close attention to developers' preferences with regard to factors such as site location, land use, and price, which might affect the profitability of the product. In order to accommodate developers' requirements, negotiations with them preceded the announcement of renewal plans even to local public officials, much less to neighborhoods in which sites would be cleared. Insulating agency-developer negotiations from the public prevented critical review by the communities that would be impacted; it also allowed the agency to claim that modifications in its plans "would compromise delicate negotiations" with investors (Friedland, 1982, p. 104).

The administrative insulation of powerful programs such as urban renewal from public scrutiny and control typically gave additional influence to already influential constituencies such as developers, who thus privately could shape agency policies to their own benefit. By contrast, less influential members of the community such as minority groups and the poor, who have often paid the social costs of renewal, have tended to have easier access to local institutions—such as public housing agencies and city councils—that possess relatively little power to affect renewal program activities.

In general, how government and its programs are organized results in different avenues of program access to different groups interested in the development and implementation of any given policy. This is of particular importance in the context of bureaucratic city government, where policy formation and execution both depend heavily on the pattern of connections between interest groups and public agencies. Like a "political lens," the "structure of government crystallizes the organization of power" (Friedland, 1982, p. 104) in the community, concentrating and magnifying the power of already influential, well-connected groups while diffusing the relatively small power of the less influential with regard to how policies actually are translated into practice in the local environment.

Such has been the case with postwar urban redevelopment, largely conducted as a program fostering the interests of corporate business and other large-scale urban institutions. Structuring the urban renewal agency so that it has been independent of other local public agencies (yet strongly dependent on private developers) has tended to ensure its responsiveness to corporate and institutional interests, in contrast with a lack of responsiveness to lower class and minority groups dependent on public sector protection of their interests. Chapter 6 examines some of the attempts by those groups to achieve greater control over urban programs relevant to their own communities.

NOTES

1. This transition has not been uniform across different industrial era cities. It has been most difficult and least complete for cities whose economies were most narrowly "smokestack" industrial in composition (e.g., cities such as Akron, Ohio; Gary, Indiana; and Buffalo, New York). It has been most successful in cities that already had well-developed centers of corporate office, commercial, and related service activities (e.g., New York, Chicago, Boston, and San Francisco); for related discussion see Mollenkopf (1983, p. 40).

2. Similar prorenewal political alliances were formed in numerous older central cities, largely during the first three decades after World War II. Some of the better known examples include the Greater Baltimore Committee, the New Boston Committee, the Chicago 21 Project, the Greater Philadelphia Movement, and Pittsburgh's Allegheny Conference on Community Development (Frieden & Sagalyn, 1989; Judd, 1988). At the national level, organizations such as the Committee for Economic Development and the Urban Land Institute (ULI) worked to enlist federal support for programs of postwar economic growth and urban development, partly to assure against a postwar Depression and partly to help save older central cities from the effects of more than a decade's neglect, injurious to the cities' economies and also sufficiently

widespread to "eventually hamper national economic growth" (Fox, 1986, p. 95). The ULI was established in 1939 by the National Association of Real Estate Boards (NAREB) as a formally independent research and advisory organization, particularly to serve cities interested in redeveloping their declining central business districts. Representing major central city property owners and developers, ULI aimed to protect existing downtown property and the interests of related financial institutions (banks, savings and loans, and insurance companies) and to facilitate new CBD construction through publicly assisted redevelopment.

To this end the institute "developed a legislative proposal for a separate federal agency," through which it hoped to arrange subsidies to cities to cover "most of the difference between the costs of site acquisition and clearance and the resale price the city received for private developers," mainly for downtown renewal properties. Following passage of the 1949 Housing Act, ULI provided consulting teams including "corporate, financial and real estate executives" to assist local redevelopment sponsors ("usually corporate organizations") in addressing planning and financial issues in the redevelopment process (Friedland, 1982, pp. 87, 88).

Overall, ULI played a critical role in linking downtown real estate and corporate interests and in lobbying for federal subsidy to renewal. In the latter, it was joined by the U.S. Chamber of Commerce, which acted as an important mechanism for achieving broad business consensus spanning small local firms to multinational corporations and as a "transmission belt" for urban policies developed by other business organizations concerned with urban redevelopment, such as ULI and NAREB. These national redevelopment-concerned organizations played key roles in linking older cities into "a network which helped catalyze local action on federal Urban Renewal" (Friedland, 1982, p. 90), influencing during several decades both local corporate progrowth strategies and the development of related federal urban policies. In the position of key renewal network operators endowed with major corporate resources, it was possible for such downtown business-oriented lobbying organizations increasingly to convert the 1949 Housing Act into a vehicle of CBD renewal, outmaneuvering the advocates of affordable housing, who represented a potentially key segment of any significant opposition coalition, and minimizing subsidies for public housing through extensive antipublic housing propaganda and political campaigns (Gelfand, 1975).

[6]

Cooperative Federalism
and the War on Poverty

ANOTHER APPROACH TO
RENEWING THE URBAN COMMUNITY

Background to the Declaration of War on Poverty

During the first generation after World War II, older large cities experienced a period of significant change along two major dimensions. On the first dimension—economic—there occurred a postindustrial shift from an industrially based economy to one emphasizing services and administration. This resulted in major changes in the urban occupational structure and a serious loss of economic opportunity for large segments of the industrial working class, particularly among minorities. Change on the second dimension—demographic—centered around two opposing population migrations, which together constituted a massive dual migration. Flowing inward during the war and into the postwar period was a vast immigration, particularly of uprooted rural newcomers to large northern industrial centers—attracted there first by the lure of expanding defense employment and later by the wider variety of opportunities typically associated with big cities. Flowing outward during several decades following the war was a rapid and continuing emigration of urban population and jobs, resulting in significant net loss of social and economic resources by central cities to surrounding suburbs or to other regions.

This migration of rural poor and minority populations into central cities continued opposite to the largely White middle-class out-migration of the postwar period until the recession of the mid-1970s. The influx of low-income populations was based largely on the same promise of urban economic opportunity that had benefited earlier generations of immigrants over the preceding century. Yet despite that inviting promise, as of the mid-1960s— amid a period of economic expansion and low unemployment—the immigrant populations still were not well integrated into city economies. The problem was most severe for Black immigrants, who were disadvantaged in the search for employment not only by low industrial skills but also by housing, educational, and job discrimination. In 1940 Black unemployment had been just 20% higher than that of Whites, but by 1953 it was over 70% higher, and by 1963 it was more than double White unemployment (112% higher). These increases in unemployment coincided with the migration of almost 1.5 million southern Blacks into northern industrial cities in the 1950s (Shank & Conant, 1975, p. 278).

FEDERAL RESPONSE TO
THE REDISCOVERY OF POVERTY

Amid preoccupation with World War II and the suburbanization boom of the 1950s that followed it, earlier issues of poverty tended to fade from public concern (Wilson, 1987). Although the economic problems of the urban poor already were serious by the mid-1950s, there was no official notice taken of them under the conservative Eisenhower administration, which held office until 1960. Indeed there was not even an established official standard for measuring poverty, and poverty did not become an officially recognized public issue until the early 1960s under the Kennedy administration. Various observers stress that this occurred in response to a number of politically relevant circumstances. First was the expanding movement for civil rights of non-White minorities, which began as a movement of protest against legal and political discrimination in the South (Piven & Cloward, 1977), later broadening to include economic discrimination in the industrial cities of the North. Second was the emerging evidence of growing levels of unemployment and poverty in depressed areas, concentrated particularly in older central cities (Harrington, 1962). Finally, there was the recognition in the Kennedy administration of the growing significance of the Black vote, which provided the narrow margin of victory to the Democrats in the 1960 presidential election.

Incorporation of urban Blacks as a Democratic voting bloc came to be regarded as vital, considering the heavy out-migration from central cities of large numbers of the Democratic urban constituencies built up during the New Deal, which included liberal professionals as well as unionized industrial workers. In view of the loss of these largely White middle-class constituencies to suburban locales and uncertainty as to the impact of suburban life on their political loyalties, there seemed good reason to integrate the incoming urban migrants both economically and politically.

On the economic side, the planners of the War on Poverty aimed at expanding the structure of social policy beyond the basic income security programs that had been established by the New Deal. The new antipoverty programs would emphasize job preparation for those having difficulty gaining entry to the workforce, rather than governmental transfer payments such as Social Security, largely targeted to retired workers.[1] They would be designed to respond particularly to needs of young low-income and minority persons, who experienced disproportionately high levels of long-term unemployment and high rates of poverty, but who were not covered adequately by existing income-maintenance programs based on transfer payments.

Politically, it was hoped that antipoverty programs would contribute to rebuilding a broad base of Democratic electoral support, particularly in older big cities (Piven, 1974a). Assistance from these federal programs could bind the urban newcomers to the national Democratic party in a way that the local Democratic machines would have found difficult to match, because those organizations no longer commanded the resources they once had enjoyed. Moreover, typically dominated by White ethnic groups, the remaining machines lacked much motivation to organize and work with the new minorities. It was in the context of these types of economic and political concerns that the War on Poverty was developed by policy strategists working with the Democratic administrations of the 1960s.[2]

In the summer of 1963 President Kennedy indicated that he was interested in a pilot program to combat poverty in a limited sample of about 10 cities. Kennedy's own concept of an antipoverty program envisioned a strategy of government action for job creation together with services that would help the unemployed link up with jobs. His projected legislative program for 1964 (put together shortly before his assassination in November 1963) included a tax cut to revive the economy and expand employment opportunities; another part provided for manpower training aimed at jobless persons who needed new skills for a changing job market, plus job counseling and health care (Shank & Conant, 1975). In short, the Kennedy program rejected the notion that governmental stimulation of the overall economy would be sufficient by itself to help those trapped in socially structured pockets of poverty at the bottom of it. Hence the inclusion of a package of manpower categorical grants, which were targeted specifically at improving the employment readiness of those caught in conditions of structural unemployment associated with low-income or minority status, such as low skills or poor health.

In November 1964, former Vice President Lyndon Johnson, who had assumed the presidency on Kennedy's death, was elected president by a landslide majority of more than 60%. Johnson's electoral mandate opened the way for an expanded poverty program, which he pursued with vigor and impatience. This was consistent with Johnson's style of leadership and with his background as a committed New Deal Democrat. Within his first year in office, this well-connected political entrepreneur (with a long background of leadership in the Congress) had engineered the passage of the Civil Rights Act of 1964 (protecting the voting rights of minorities), as well as the legislation underlying the War on Poverty.

The War on Poverty:
Some Underlying Assumptions

The War on Poverty was officially launched in August 1964 with the signing of the Economic Opportunity Act by President Johnson, establishing an Office of Economic Opportunity (OEO) directly within the executive office of the president, reflecting its importance to him at the time. Under its provisions, the act included a mix of manpower development and educational programs, plus community organizational, legal, and health programs, which together sought to deal with the different facets of the problem of poverty (Sundquist, 1969). To the extent that there was a single unifying theme, it was one of expanding opportunity at both the individual and community level so that poor people could compete more effectively with others in the economy (and in political life as well).

One underlying assumption that influenced the design of the War on Poverty concerned the issue of large-scale long-term unemployment, often identified as a significant cause of poverty. Supposedly, in the "full-employment" economy of the mid-1960s, unemployment was no longer due to problems of how the American economy performed, thanks to the development of advanced techniques of public economic management. The notion was that with the help of appropriate governmental manipulation (as in the use of tax cuts to stimulate business investment), the economy essentially was capable of producing the jobs needed to absorb the unemployed.[3] Hence the key to addressing the problem of long-term unemployment and thus reducing poverty was identified with remedying the defects of the poor themselves—in education, job skills, and attitudes toward work—in order to improve their access to jobs, rather than changing the structure of the economy or the mechanism by which jobs were provided (as would have been the case with a broad expansion of public sector employment, for example). In this perspective (Warren, 1977), the major problems to be attacked were the educational and attitudinal deficiencies of the poor, typically associated with the conditions of "cultural deprivation," in which persons living marginally to the mainstream culture lack either employable skills or norms of hard work and achievement.

Beginning from such assumptions, fighting poverty would emphasize a strategy of governmentally assisted expansion of social services in order to overcome the educational and attitudinal shortcomings of the poor. Thus publicly funded social services would act as the functional linkage between the unemployed poor and a robust, expanding economy. The traditional social service focus on individual casework, psychiatric services, and family counseling now would be broadened to include basic educational services and job training and placement. During the 1950s this outlook grew in influence. However, by 1960 a distinctive new approach also had developed based on a theory of the relation of social deviance to problems in the community's structure of opportunity (Cloward & Ohlin, 1960). This theory underwent preliminary testing through demonstration projects in New York and several other big cities, funded by the Ford Foundation and subsequently the President's Committee on Juvenile Delinquency in the early 1960s (Piven, 1974b). By the advent of the War on Poverty in 1964, this approach had taken the form of a new type of antipoverty strategy that came to be known as *community action*.

Within its first year of operation the Community Action Program (CAP) emerged as the particular aspect of the War on Poverty that suggested the sharpest departure from the social services approach. In significant respects, community action represented an attempt to go beyond simply improving the poor as separate individuals. It sought to organize and empower the poor as residents of poverty neighborhoods in order to press local institutions (such as public schools and employment agencies) to change established practices as necessary and actively engage in assisting them in gaining access to new economic opportunities. Thus Community Action could be viewed as an organizational change approach to providing linkage between the community's poor and the expanding economy.

At the core of the Community Action approach could be found a distinctive assumption: that the condition of the poor cannot be explained fully or improved by focusing only on their individual characteristics and defects but also requires an understanding of them in the context of the local institutions and communities by which their lives and opportunities are typically bounded. Only by changing the operation of these institutions and the relationship of the poor to them could the structure of available opportunities be changed (Katz, 1989). This was clearly a different approach from the strategy of improving the poor by changing their personal attitudes or reducing their individual cultural deprivation.

However, this new and relatively untested community-based approach to opening up local economic opportunity by changing the community's institutionalized opportunity structure was able only to challenge but not to displace the more conventional approach that was rooted in traditional social service practice. Thus the two approaches were incorporated into the antipoverty efforts of the period. In short, a variety of social service programs were brought together in the War on Poverty, responding to the concerns of the cultural perspective; alongside these and administratively intertwined with them the Community Action Program was developed, responding to issues of structural change. Hence the central theme of expanding opportunity came to be pursued through a programmatic combination of the two perspectives discussed. I will describe briefly both types of program and then examine how they were combined under the umbrella of Community Action.

Social Service Programs

Generally the social service programs of the War on Poverty reflected a strategy that sought to upgrade the poor—especially the young poor—through basic education, job training, or work experience that would help them become more productive and improve their economic opportunities. As categorical grant programs, they were targeted to specific categories of need, related to particular clienteles in the poverty area population. For preschoolers there was the Head Start program, which provided several years of early educational socialization for prekindergarten children. For low-income teens there were Neighborhood Youth Corps (part-time and summer work programs), Job Corps (2 years of job training in skills centers located away from home neighborhoods), and Upward Bound (college preparation, usually by way of summer sessions on college campuses). For adults, there were basic literacy and vocational education programs.

These service programs ranged from the relatively conventional Job Corps (modeled to some degree on earlier New Deal job programs) to more innovative programs such as Head Start and Upward Bound (Henig, 1985). In addition, there were legal aid and health services programs to assist in removing other barriers to employment of minorities and the poor.

A number of other important social programs came to be associated with the War on Poverty after 1964. Foremost among these from the perspective of this book was the Model Cities program, a specifically urban sequel to the Community Action Program discussed in detail later in this chapter. Other programs with substantial urban impact included Title I of the Elementary and Secondary Education Act (providing federal aid for education of disadvantaged children), Aid to Families With Dependent Children (AFDC), food stamps, and the Housing Act of 1968—each having relevance for improving the quality of life and future opportunities of inner-city populations (Henig, 1985; Katz, 1989).

THE COMMUNITY ACTION PROGRAM

The Community Action Program (CAP) was the most unique aspect of the War on Poverty in that it was designed to be actively experimental and innovative in its response to community problems. It was to be concerned not only with improving the skills and capacities of the poor as individuals but also with their relationships to key social institutions that shaped the economic and political context of their life in the community. This gave an organizational change aspect to CAP that took it beyond conventional social service programs to the possibility of mobilizing communities for significant change in their local institutions (Vanecko, 1969).

Evolution of Community Action Objectives

The Community Action Program was charged with an ambitious mission involving three general objectives that were not always consistent with one another: improving social services, providing full coordination of services, and encouraging resident participation. First, the program would act to increase the amount and upgrade the quality of social services aimed at designated poverty areas. The intention was that each local program agency would be responsible for designing and funding services suitable for its own locality. Services that the agency did not itself develop it would purchase from local public agencies or nonprofit providers. Second, the agency would seek to facilitate the coordination of all public and private resources relevant to antipoverty efforts in designated areas. Here it would play the role delegated to it by its federal parent agency (OEO) of local organizational coordinator for the delivery of antipoverty services to targeted areas. Third, it would work to encourage and facilitate the *maximum feasible participation* of residents of targeted poverty areas in developing and implementing the Community Action Program (Kravitz, 1969).

Of these three aspects of the program, the two that generally were most distinctive as well as problematic were the attempt at comprehensive coordination of local services and program support for participation of residents who would be affected by program implementation. The coordination function of CAP was intended to serve as a means of combining traditionally uncoordinated social service programs (educational, job training, health, legal, etc.) produced by already existing local agencies and directing them together with its own newly developed programs to designated poverty areas. Specifically, this was to be achieved through establishing a *community action agency* (CAA) in each of the approximately 1000 communities that received program funding. After studying local poverty conditions, CAA staff would identify "the most severe pockets of need" and develop "a program for these areas that would affect all relevant institutions"—such as "the schools, social services, job opportunities" (Kravitz, 1969, p. 60)—so that the various program facets would contribute in an integrated manner to reducing poverty. Ideally, each CAA was to act as an organizational mechanism for programming and coordinating an appropriate mix of antipoverty services at the level of the local community, concentrating them in severe poverty areas to complement and reinforce one another. Thus CAAs, as organizations specially created to administer local delivery of program services, also were intended to play a significant coordinative role in a multifaceted strategy against poverty. These were high expectations for total newcomers in the complex, competitive local organizational environments particularly characteristic of large cities.

Like the local housing authorities of the public housing program and the local renewal authorities of the urban renewal program, which acted as managers and coordinators of projects related to their respective programs, CAAs were federally delegated local authorities designated to act as community decision organizations in relation to local antipoverty efforts. But unlike those other programs, which typically were managed at the local level by recognized community elites, there was an aspect of program implementation unique to CAP that led to special complications. This rather distinctive aspect of the program was the notion that local residents should take part in program administration and development. The Community Action Program's stated objective of maximum feasible participation suggested the possibility of important openings to local residents in shaping the ongoing operation of the program.

This somewhat vaguely defined provision of the Economic Opportunity Act was interpreted variously by those interested in CAP. The more liberal interpretations suggested that CAAs should encourage increased levels of involvement by poverty area residents both in formal decision roles within the CAA and in antipoverty organizational and advocacy activities in the community to assure influence in CAA decision making and some degree of community control of other relevant local service institutions (Peterson & Greenstone, 1977). Just what balance should be struck between resident participation (with its connotations of resident confrontation, conflict, and control of local service-providing agencies) and service coordination (with its emphasis on negotiated consensus between the CAA and other local service institutions) was not specified because neither OEO nor the Congress indicated which of these basic program goals had greater priority. Thus it was left in doubt in which direction CAAs might go; specifically, whether CAAs "were to be primarily coordinators of federal programs and operators of OEO programs, or organizers of and advocates for the poor" (Shank & Conant, 1975, p. 289).[4]

During the first stage of the Community Action Program, lasting about a year, antipoverty managers were most concerned with expanding the volume of social services and improving their coordination in targeted areas. However, at the federal level, powerful established agencies such as the Department of Health, Education, and Welfare were resistant to the prospect of a new agency such as OEO coming on the scene and directing them in the coordination of their programs. Lacking the organizational clout and experience necessary to influence significantly the program activities of regular cabinet agencies such as HEW or the Department of Labor, OEO's leadership "never seriously tried to establish a coordinating function for OEO" at the federal level (Peterson & Greenstone, 1977, p. 245).

Thus coordination of services in CAP fell by default to the local level of program implementation, specifically to the CAAs. Reinforcing this, the notion of delegating to the CAAs the responsibility for coordinating local antipoverty services remained attractive to the powerful Bureau of the Budget, particularly because the bureau did not propose to spend a great deal of money on poverty programs. Its major objective was to find instead a new means by which to deliver existing services with greater efficiency (Moynihan,

1969). Consequently, despite their lack of previous planning experience, the highly complex task of coordinating the diverse elements of the War on Poverty was left to the CAAs.[5]

In this way, the CAA came to be designated as the officially recognized community-level organization for translating the broad antipoverty program objectives—service expansion, service coordination, and resident participation—into specific local activities. CAA was to be the agency through which funds would be channeled from OEO, the federal parent agency, to the particular combination of program efforts fielded in the local community. Ideally, the basic approach was that "local initiative" programs would be devised by each CAA in response to specific local conditions. However, many newly established CAAs lacked the organizational capacity or experience to research, develop, and prioritize local initiative programs. Hence CAP funds were used to pay for local implementation of relatively standardized "national emphasis" program designs, which were developed at the federal level by OEO planners to meet broadly identified essential community needs such as the Head Start program and several others, including neighborhood health centers, family planning, and legal services (Sundquist & Davis, 1969). As CAP evolved, these prepackaged programs grew in importance, receiving increasing proportions of annual OEO funding. Head Start was the first and most important of the national emphasis programs, receiving 38% of all CAP funds during the Johnson administration from 1965 through 1968. Taken together, Head Start and about another half dozen national emphasis programs received approximately 50% of CAP funding (rising from 40.8% in 1965 to 53.2% in 1968). During the same period, local initiative programs received only about one third of CAP funding (Levitan, 1969, p. 123).

Specific program implementation—whether of locally initiated program designs or of national emphasis designs—could be arranged by CAAs through contracts with public agencies and nonprofit service-providing agencies (e.g., local public schools, state employment agencies, and private social service agencies) as called for in the given program design. This could require modifications in the practices of these local agencies based on adjustments negotiated between CAAs and the different agencies involved. In 1966, to reinforce the CAAs' coordinating influence, all other federally funded poverty-related programs in CAP localities (such as compensatory education and public housing) were made subject to review by CAAs as to their ongoing consistency with antipoverty objectives (Sundquist & Davis, 1969); properly utilized, such review could provide a basis for negotiating further modifications in local agencies' programs. Thus the process of arranging for delivery of services in the community involved continuing negotiation with diverse service organizations at the community level, in addition to communication with OEO at the federal level regarding the national emphasis programs. Furthermore, in order to facilitate ready access for local residents to the particular programs that came to be fielded in target areas, CAAs often established neighborhood offices that served as "one-stop centers" to help consolidate the delivery of various programs, as well as for resident organization activities and provision of program outreach and information to residents (Peterson & Greenstone, 1977, p. 245).

These related arrangements placed CAAs amid an intricate web of interorganizational relationships involving providers of traditional social services and of new locally initiated as well as federally designed services. Through careful allocation of program funds and vigorous pursuit of outreach activities, CAAs might at least succeed in expanding the volume of antipoverty services delivered to the target locality. However, it soon became clear that the more complex goal of improved coordination of services would be considerably less attainable. With so many competing organizational interests among diverse public and private service agencies and so little preparation for the task of effectively combining their efforts, service coordination tended to become reduced in priority as a program goal. In addition, those CAAs that took seriously the charge of maximum feasible participation and that aggressively organized local residents often came to be shunned as not behaving in a "responsible" manner in the organizational context of the community. On the other hand, more moderate CAAs (often found in smaller communities) were viewed as little more than "another specialized organization quietly administering a few programs designed in Washington" (Sundquist & Davis, 1969, p. 46).

In either case, the image of the CAA was not consistent with the comprehensive service coordination function envisioned for it by OEO or the Bureau of the Budget; over time, these impressions tended to harden into a stereotype of irresponsibility and limited competence that denied legitimacy to the CAA's local coordinating role. As a result, the program objective of service coordination came to be generally deemphasized so that CAAs increasingly came to be identified either with service expansion, with local resident participation, or with some mix of the two.

Citizen Participation:
CAAs and the Community Organizational Field

In this regard, although expansion of social services was viewed as basic to reducing economic poverty in targeted areas, citizen participation was seen as that aspect of Community Action that aimed at reducing local "political poverty" (Peterson & Greenstone, 1977, p. 241). In a society whose pluralist political values encourage the notion that the system is, or at least should be, open to participation by all groups, political empowerment of the poor certainly would seem a worthy goal. Yet the participatory aspect of Community Action often led to sharp controversy, particularly in those communities where it was most seriously pursued. In this section I examine some of the reasons for such controversy.

Citizen participation in urban programs was first introduced in the Housing Act of 1954, which amended the original urban renewal legislation of 1949. This act provided for the creation of city-level citizens' advisory committees, as an element of a "workable program" that set forth explicit regulations for the local conduct of an improved renewal program—one presumably more considerate of public response and sensitive to relocation needs. For the first time in federal legislation, citizen participation was identified as one of the elements required as a precondition for the city receipt of federal program funds. In practice, the elite blue-ribbon citizens' advisory committees to the renewal program generally consisted of leading local professionals and businesspeople (Gilbert & Specht, 1977) and were intended more as a basis for legitimating the renewal program than for seriously modifying it; ordinary local residents were given essentially no part in the key renewal decisions affecting their neighborhoods.

Nevertheless, the 1954 Housing Act's provisions for citizen participation in urban renewal have been pointed to as an early step on the road to "bureaucratic enfranchisement" (Fainstein et al., 1983, p. 21)—a precedent through which government agencies first opened the way to some form of participation by local residents in urban program administration. Although citizen participation began as a strategy of co-optation, providing legitimacy for renewal policies that were often detrimental to urban poverty neighborhoods, a decade later it had become an instrument for organization and possible empowerment of the residents of poverty areas. Equally important, the CAAs provided for the first time a structured bargaining arena (Pressman, 1985) in which residents regularly could confront public and private decision makers. Directed to this arena, vigorous citizen participation could make it possible for local residents to negotiate successfully for concessions to local community needs.

Thus a decade after the 1954 Housing Act gave urban renewal a fresh start through its workable program, Community Action was projected as a framework for a much more substantial type of citizen participation. Using the

CAAs, existing institutions in poverty areas were to be opened to the scrutiny of neighborhood residents and made more responsive to the often neglected needs of the poor. Consistent with the notions of "participatory democracy" and "grassroots democracy" that had caught the imagination of significant segments of the public, including various community activists and social planners (Altshuler, 1970, pp. 13-14; Perlman, 1979), resident area representatives would be given an active part in program planning and allocation of resources in relation to priorities that residents played a significant part in defining.

In this respect, although CAP included antipoverty programs that essentially concentrated on providing direct services for the poor, it sought to go significantly beyond this. After more than a decade of elite-controlled urban policy focused on physical renewal of central business districts, CAP held out a double possibility. First, it might act as a vehicle for introducing a missing dimension to urban redevelopment policy in the form of programs for social rather than physical renewal, targeted particularly to the residential slum areas surrounding the CBDs. Second, such programs would not be operated through conventional top-down management by remote bureaucratic agencies. Instead in significant ways they would be grassroots programs rooted in local citizen organization and participation, including organized bottom-up advocacy by poverty area residents concerning local community needs. The Community Action Program would serve as the institutionalized means by which poverty area problems and issues could be identified, aggregated, and authoritatively represented in the larger urban community.

Because such a mechanism never before had existed in American communities, it opened an unprecedented opportunity to exert pressure for significant institutional change at the local community level. The intention was to use the CAA not only to upgrade local services to families and individuals but also to encourage citizen participation in pressing for changes in the decision structure of the community in locally relevant institutions such as the schools, public employment agencies, welfare departments, and ultimately city government. This role of the CAA as an instrument for reforming local government and related public service bureaucracies on behalf of expanding economic and political opportunities for the poor proved to be the single most controversial feature of the Community Action Program.

In addition, a related feature of CAP was that it required its implementing organizations, the CAAs, to include representatives of the poor and provide them with decision-making roles regarding program design and implementation. It was left to the particular mixture of interests and perspectives incorporated in each CAA to determine the specific nature and extent of those roles. The governing board of each CAA would not be confined to community elites as traditionally had been the case with boards of urban social service agencies or with urban renewal advisory committees. Instead it was conceived of as a "three legged stool" with representation from the public sector (public officials and city agency representatives) and the private sector (representatives of private social service agencies and business and professionally based civic leaders); for the first time, these community leadership elites would be joined by elected representatives of targeted poverty areas (Wofford, 1969, p. 83).

Initially, CAP planners expected at least a basic degree of cooperation among these different sectors of the community on behalf of improved service coordination. But such expectations typically were not realized. Owing to their diverse backgrounds, CAA board members represented quite different interests and perspectives, and they came to their interaction with very different levels of administrative and planning experience. Skeptical of the innovative ideas or planning competence of resident representatives, city agencies often effectively blocked the CAAs' attempts to coordinate their activities and possibly change their bureaucratic routines. As a former CAA director observed, "The local agencies weren't on the boards to coordinate but to protect their vested interests" (Sundquist & Davis, 1969, p. 75). Consequently, rather than becoming central community decision organizations coordinating local and other services for antipoverty objectives, many CAAs became a parallel set of institutions to the established local service agencies, focused primarily on delivering federally provided antipoverty services such as the Head Start program. Citizen participation in these generally moderate CAAs was largely internal to the agency, concerned with activities in the various special committees of the CAA; to the extent that such participation related to other agencies of the community, the approach amounted to conventional lobbying for change in those agencies' priorities or procedures.

In contrast, CAAs that were dominated by militant local resident organizations placed first priority on encouraging vigorous citizen participation as a means to political empowerment of poverty area residents. Although militant-dominated CAAs constituted only about one third of all CAAs in the Community Action Program, they nevertheless had important consequences for the public's general image of the program (Shank & Conant, 1975, p. 297). The object of the militants was to enable local residents to participate in shaping and implementing antipoverty policy both in relation to specific CAA programs and in the broader context of the community's governmental process as it affected all locally fielded service programs that had relevance to the status of the poor in the community. Such participation would have been taken for granted for middle-class homeowners, for example, in regard to their interests in protecting and improving their residential areas and their political-economic status as citizens of the community. However, based on the widespread image of CAAs (whether militant-dominated or not) as disruptive to the public order, there was often deep resistance to the antipoverty program (from both public service agencies and elected officials) and to its provisions for participation of the poor in program implementation and possible expansion of their political influence into the broader organizational field of the community.

In addition, there were ideologically fostered suspicions concerning what was advertised broadly as the War on Poverty's tendency to "dangerously centralize Federal controls and bypass effective state, local and private programs" (Sundquist & Davis, 1969, p. 48). This tended to create additional difficulties for program coordination beyond those involved in the sheer complexity of the task, seriously hampering the CAA's relations with other agencies in the community and with the local political establishment.

Poverty area militants often mobilized resident participation that expressed criticism of the shortcomings of established service organization programs and practices in opening new opportunities to the poor. Typically this involved challenges to local service delivery agencies and to political regimes with which those agencies shared an established organizational *domain consensus*. This was reflected in public confrontations, demonstrations, and other forms of interorganizational conflict and turbulence sparked by militant actions. These challenges to the established community power structure met with a certain degree of success in winning changes in local institutions, as well as increased services (Peterson & Greenstone, 1977; Vanecko, 1969).

However, these challenges also stirred strongly defensive reactions with significant consequences for CAA interactions and relationships. On CAA boards they heightened resistance from representatives of both public and private sector agencies to establishing cooperative relations with poverty area representatives and worked against the development of mutually accepted planning procedures required for effective program coordination. The more militant the resident organizations associated with any given CAA, the lower the likelihood of cooperation from other organizations in the interorganizational field of the community's service system. Consequently, there tended to be an inverse relationship between aggressive political mobilization of local residents and the likelihood of successful service coordination (Levitan, 1969).

COMMUNITY ACTION AND
THE COMMUNITY ORGANIZATIONAL FIELD

In general, this inverse relationship reflected the tendency of the organizational field of the community to protect its established structures of jurisdiction and competence, that is, its patterns of organizational domain, from unaccustomed challenge by organizational newcomers to the field. Organizations tend to establish domains through a process of negotiating their functional relationship to other organizations in the community's interorganizational field. Establishing a domain is based on achieving consensus in the interorganizational field regarding the "problem area covered," the "population served," and the "services offered" by the given organization (Rose, 1977, p. 471).

For purposes herein, the *social service system* of the community can be defined as that segment of the interorganizational field that relates to education, employment services, and the special needs of those in poverty, including public assistance, health, and legal services. The established social service system of the community is not a product of formal planning; rather it emerges over time through a process of mutual adjustment based on negotiated interdependencies between organizations (Levine & White, 1961). The intrusion of a new organization into this interorganizational subfield creates "a crisis for existing organizations and for the domain consensus negotiated over time." This is particularly the case "when a new organization brings with it an unclear set of domain expectations . . . and undetermined methods of organizational interaction" (Rose, 1977, pp. 472-473). Such was the situation with the CAAs.

Creation of a community action agency threatened both the organizational autonomy of each organization already in the field and the general domain consensus. The CAA's declared intention was to examine the organizational objectives and practices of established organizations and the degree to which their services actually worked to bring poor people out of poverty. The result was considerable uncertainty for the organizations of the community's service system. It was not clear just what changes would have to be made in organizational domains in order to better coordinate services offered by participants in the community's social services system or how the concept of maximum feasible participation would affect service organizational domains and routines.

An example of the reactions of organizations whose domain and operations CAAs were trying to change can be seen in the case of the public employment services. From the perspective of the CAAs, a major priority of the poor, who were their primary clientele, was to obtain jobs. Consequently, the CAAs sought to promote outreach to the poor by the established local public employment services (typically financed by the U.S. employment service and administered by the states). What they found, however, was a predisposition by the employment services to serve the employers as their primary clientele by connecting employers with the skilled and experienced personnel they preferred, rather than unemployed poor persons. As a result, many CAAs soon became involved in "the creation of competing machinery for contacting employers, for placement, and even for training—and, in doing so, inevitably antagonized the employment services" (Sundquist & Davis, 1969, pp. 49-50).

In cities with pools of unemployed too large for CAAs to service alone, private employment agencies sprang up in response to contracts for purchase of their services by CAAs and related federal agencies (such as OEO and the Labor Department). Thus, even as some of the public employment services began to develop outreach functions, they found that rather than getting the necessary support themselves, federal funding was flowing through CAAs to competitive private agencies. Public employment service officials complained that the new private agencies were inexperienced and inept, turning to the established public agencies for help in performing even basic operations such as client screening, and that the programs they had initiated were merely training the poor for dead-end jobs such as janitorial help.

Similar complaints were heard from welfare officials when CAAs undertook vigorous "case finding" outreach activities to bring in previously unreached families to welfare agencies. Although some welfare officials admitted that their outreach was limited, they blamed it on local fiscal restraints; others declared that once they received direct federal funding under the Economic Opportunity Act, they were able to develop quite creditable new training and employment services for adult welfare clients. Overall, many public agency officials spoke resentfully of the tendency of CAP and its local CAAs to bypass them and not draw on their experience, thus expressing their sense of a threat from CAAs to established organizational domains and domain definitions (Sundquist & Davis, 1969).

These reactions to CAAs were not unique to public employment services and welfare agencies. Similar concerns and resentments were forthcoming from public school and public health officials as well. But this might have been expected. The planners of the Community Action Program made it clear that they lacked confidence in the established service agencies to carry out a War on Poverty. This was why they rejected the federal Department of Health, Education, and Welfare's initial bid in 1964 to oversee the conduct and coordination of antipoverty programs. Instead they turned to creating OEO as an independent federal agency explicitly devoted to funding antipoverty programs and providing for their local coordination through the CAAs. However, by introducing the CAA as a new institution that would fund competing programs to those of the established social service agencies, OEO destined it to being regarded as a domain competitor at just the same time that it was being fielded as the coordinator of organizational domains relevant to the antipoverty effort.

In addition, the CAAs' mission of supporting the organization of poverty area residents as potential critics of local service and political institutions underlined their image as challengers of the existing domain consensus instead of efficient coordinators of community services. Although CAAs themselves did not engage in organizing activities, through their neighborhood centers they tended to be identified with groups involved in local organizing. Although most of the resident organizations they supported were headed by moderates mainly concerned with operating service programs, in the perspective of local service providers CAAs were seen as advocates of conflict with established institutions. Probably what made CAAs seem particularly threatening was their special relationship to the resources of the federal government—the array of special programs, funding, authority, and sanctions made available by the Johnson administration's War on Poverty. Black civil rights groups and Hispanic political groups both had preceded CAAs in creating organizations to develop the political power of the minority poor, but they never had resources for stimulating local change comparable to those of the CAAs. Acting with OEO support, CAAs at least initially appeared to have the potential to bring about changes in the status of the poor, empowering them for the first time as an effective interest group and creating the possibility of a shift in the balance of power in the urban community (Sundquist & Davis, 1969).

As events transpired, CAAs had only a limited time after their creation in mid-1964 in which to freely pursue their threefold mission, for in the next few years that mission was to be seriously challenged and curtailed. Only a year later at a meeting of the U.S. Conference of Mayors in June 1965, strong objections were raised concerning the CAAs' role in supporting citizen participation of the poor, with several mayors accusing OEO of undermining local government by encouraging class struggle. By 1966 there were signs of erosion of support for the citizen participation thrust of the early Community Action Program inside the Johnson administration, as well as among big-city mayors and other critics. Representations by mayors opposed to militant resident participation led the administration to reinterpret maximum feasible participation as a guideline for giving jobs to the poor in local antipoverty programs, rather than encouraging their involvement in the planning of such programs. In addition, Congress reduced the appropriation for CAA local initiatives activities, thereby limiting the role of resident participation in program development. Most funds for CAAs would be earmarked now for Head Start and for programs initiated by Congress, particularly for new employment or social services programs.

By 1967 many CAAs had begun to commit themselves largely to a service delivery role; only among the relatively few CAAs in which residents were dominant on the governing board was the emphasis still on participation and political action for institutional change. However, the urban riots of 1967 provided political ammunition to a growing conservative opposition, which worked to tie responsibility for those riots to militant resident activism. As a result the House of Representatives passed an amendment that year to the Economic Opportunity Act that gave new powers to local government with respect to CAAs. This amendment strengthened local government influence by reserving one third of the seats on the CAA board to public officials and allowing up to one third to be held by representatives of major private interests such as business, professional, labor, religious, or social service groups. By voting together, these established public and private organizational interests could determine CAA policy, shifting its activities away from politically threatening participation to service delivery. In addition, city hall now had the authority to establish a CAA as one of its own municipal agencies or to designate an agreeable private organization of its own choosing (Levitan, 1969).

In spite of this congressional backlash, CAA workers had demonstrated that they could serve as an important organizational resource, acting as a bridge between local dissidents and city hall, and that they actually had contributed to minimizing local involvement in urban riots. By 1968 many mayors who at one point or another were skeptical about the CAAs had come to recognize the usefulness of the (now relatively tamed) CAAs as channels of communication to inner-city neighborhoods. Consequently, although much concern had been expressed in Congress over the need to put control of the Community Action Program more firmly in the mayor's hands, less than 5% of mayors in participating communities took the option of bringing the CAA directly under their own control (Shank & Conant, 1975, pp. 296-297).

Overall, CAAs had to work against an organizational domain consensus in which two major aspects of their mission—service coordination on the one hand and resident participation on the other—were resented widely and considered illegitimate. Thus it was particularly difficult for CAAs to attempt to pursue both these goals simultaneously in the context of a skeptical, hostile local organizational environment. Consequently, most CAAs did not give them equal emphasis. Most endorsed resident participation (generally of the moderate type) along with expansion of antipoverty services as their main working objectives. This typical choice by CAAs can be seen to be rational under the constraints imposed by the local organizational field because service coordination had proven unfeasible, depending as it did on the cooperation of other community decision organizations (which generally rejected the CAA's authority in this area).

COMMUNITY ACTION AND
THE INTERGOVERNMENTAL SYSTEM

Between the 1930s and the 1960s the number and variety of categorical grants greatly increased, producing fragmentation and duplication in the federal grant-in-aid system; in addition, during that period the grant system became dependent increasingly on how local officials chose to apply the federal resources that it offered them. Thus, by the mid-1960s, the system of Cooperative Federalism had in fact become both more complex and more decentralized than it was at its origins, contrary to simplistic stereotypes that often have emphasized only the possibilities of increased centralization by Washington (Henig, 1985).[6]

The result was a system in which program outcomes at the level of local implementation had become considerably less certain, particularly with regard to redistributive policy—thus the interest in establishing CAAs as a local instrument to provide program coordination in a grant system otherwise lacking in much coherence. Hence, also, the official encouragement of participation by local residents in CAP implementation, for at least two main reasons: first, to attract poverty area residents as an active clientele of the program and thereby maintain the CAA as an organization grounded in a secure base of support (Peterson & Greenstone, 1977) and, second, to ensure that these clients received the benefits intended for them, rather than having those resources diverted to other purposes by local bureaucrats or by other local power wielders, whether northern urban machine bosses or southern politicians disinclined to channel federal benefits to local minority communities (Friedman, 1967; Moynihan, 1969).

Community Action and Creative Federalism

Like several other urban-relevant programs of the New Deal and after, Community Action was a federally funded redistributive program whose administration essentially was decentralized. Administrative control of the program was delegated by OEO to approximately 1,000 CAAs located in participating communities across the country. To encourage resident participation and program innovation, these were not even required to be city agencies but simply could be private voluntary organizations that were recognized by the city. In this respect, the grant relations involved went a step beyond New Deal Cooperative Federalism (which in public works and housing programs had bypassed the states to deal directly with city governments) to a new stage—that of Creative Federalism, in which federal funds and program authority bypassed city governments, instead going directly to a variety of nongovernmental or quasi-governmental organizations in local communities.

In this way, Creative Federalism created many "paragovernments" that could critique and compete with the programs of established local agencies, particularly in bringing services to inner-city areas. But the CAAs as paragovernments were by no means simply extensions of OEO, the central program agency for CAP, in its efforts to induce structural or procedural changes in local government agencies on behalf of improving local antipoverty services. At the same time, neither were the CAAs simply free to take action locally as they saw fit because their organizational roles placed them somewhere between the federal antipoverty program, city governments, and local poverty areas, subject to tugs and pressures from each of these organizational arenas. Thus the relationship of the CAAs both to OEO at the federal level and to city government on the local level was often shifting and uncertain. For example, although initial funding arrangements for CAP could be described as a type of targeted local-area block grant[7] allowing considerable discretion to CAAs to design new local service programs, within a few years the introduction of separately earmarked national emphasis programs produced in effect a recategorization, with those federally designed categorical programs (notably Head Start and several other programs, including legal and health services) rather than locally designed programs accounting for over half the funding to CAAs (Levitan, 1969, pp. 123-124; Wofford, 1969).

The competition for federal funding between local initiatives and national emphasis programs highlights the underlying tension between federal control of the purse strings and local control of program administration embedded in the system of Cooperative Federalism. In this case, the Community Action Program was a creative intergovernmental variant of Cooperative Federalism that stretched that system to the limit of its political possibilities. From the viewpoint of its ultimate controllers at the time (the Washington Democratic establishment that provided its funding), the Community Action version of Creative Federalism proved to involve excessive risk that acceptable limits of political relationship might be passed, for example, the risk of recipient community organizations and the CAAs on whose boards they were represented turning against local Democratic party organizations and their leadership. Thus the introduction of national emphasis grants was as much a mechanism by which federal program administrators could establish control over local program outputs as it was a means of providing a floor of basic services for targeted poverty areas at a reasonable level of quality.

In creating and funding the Community Action Program and its local agencies, Washington antipoverty planners hoped to establish an organizational mechanism that soon would produce improvements in the volume and quality of antipoverty services and in the longer term provide for good coordination (hence fiscal efficiency) of all services in the community relevant to this objective (e.g., local public schools and employment offices). Strong resident participation in the program could be helpful to attaining each of these goals, but the level of intensity and the particular strategy to be pursued by participating residents were matters not readily designed at the federal level nor easily managed at the local level.

As I have noted, resident participation of a fairly simple and moderate type (residents working as staff in the local program) could contribute to ensuring that improved levels of services actually were delivered to their appropriate targets. In the best of cases, sustained well-planned resident participation of a somewhat more sophisticated and more vigorous type probably would be required to pressure the local bureaucracies actually to improve their own services and to cooperate with CAA efforts at service coordination. Overall, the hopes of Washington program planners and administrators were realized only partially. They got substantial improvement in the volume of services, but very little in the way of improved service coordination. The role of service coordinator was one that realistically required a federative status for the CAA, in which it would have possessed authority clearly superior to individual agencies in the local organizational field for public services. But in practice, CAAs were thrust into essentially coalitional field situations in which successful coordination was dependent on voluntary cooperation by member units of the field. (See Chapter 3 for discussion of federative and coalitional field types.) Thus the CAA's service coordination function could be stymied by the lack of willingness for cooperation by any number of local agencies.

With regard to resident participation, OEO planners got something other than what they had bargained for. Despite planners' hopes for the development of consensual decision processes inside the CAAs and disciplined, well-organized lobbying with agencies outside the CAA, a limited but vocal number of CAAs engaged in strongly militant resident participation in a politics of public confrontation not only with local bureaucracies but with city halls as well. Although the CAAs involved amounted to only a small fraction of the total nationally, their tactics were seized on by opponents of the antipoverty program and by those public officials nervous about the challenge that the program posed to the local organizational status quo, as well as its potential threat to their own authority.

As a result, at the local level the reaction to militant resident organization and confrontation was strong and swift. Big-city mayors in particular (a majority of whom were themselves Democrats) complained bitterly through the U.S. Conference of Mayors to the executive branch, represented by Lyndon Johnson's vice president, Hubert Humphrey, as well as to local congressional representatives, that a program developed and funded by the Democratic-controlled national administration was stirring up public demonstrations and protests against their own local administrations (Haider, 1974). Congress responded by passing the Green amendment of 1967 to the Economic Opportunity Act, effectively limiting local citizen participation in the antipoverty program to no more than one third of the seats on the CAA board, well below the original maximum feasible participation that had opened the possibility that sufficiently aggressive and cohesive resident organizations could win majority influence over local Community Action programs (Sundquist & Davis, 1969, p. 38).

What is to be learned from this? At the least, it should be noted that making significant administrative changes—whether reforms of the urban governmental system of the early 20th century or of the urban intergovernmental policy system of the late 20th century—unavoidably will involve changes in political as well as administrative relationships and shifts in the balance of community power and in the interests served by government. Thus there can be no reform of public bureaucracy without impacts on the politics of the community, including impacts on which groups have access to public resources and to government officials, appointed and elected. Opening the bureaucracy to groups previously excluded from significant access (low-income groups and minorities, for example) held out the possibility not only of a redistribution of public material/economic benefits but also of a redistribution of political influence and of access to the policy process for deciding on the distribution of the various benefits that government can provide its citizens, such as public services, public jobs, public status, and ultimately the opportunity to continue influencing all those factors of community life on a long-term basis.

This is what was so promising about the Community Action Program for those residents in the poverty areas of American cities who became active participants in the program, not viewing themselves merely as its employees (in that sense still benefit-clients). They saw that they might become constituents of its policies and of the policies of other public institutions relevant to their communities, thus influencing the flows of public benefit over the future. However, this potential influence that was so promising to these poverty area activists often was threatening to other interests who already had achieved some influence in the existing distributions of benefits and were concerned that changes in those distributions might reduce their own future influence and benefits.

The special political conditions under which CAP was initiated, including Lyndon Johnson's landslide election, soon began to fade away against the broader background of political unrest associated with the peaking of the civil rights movement and the development of a backlash against that movement and the War on Poverty, both of which would be displaced shortly as national priorities by the country's entry into the war in Vietnam. Even before that, the turbulence associated with the more militant manifestations of Community Action and the lack of public preparation and understanding for that program produced strong reactions against it—particularly against what seemed to many in official positions to be its excessively radical conception of creating maximum feasible opportunity for the sharing of political influence and power by groups of low social-economic status, traditionally considered marginal to the existing social system and its governmental institutions.

In this strained atmosphere, it took only the volatility of a limited number of militantly activist resident groups engaging in a local politics of confrontation to produce a decline in the image of the program, first in the eyes of public agency officials on whose organizational domains it threatened to impinge and finally in the thinking of the Johnson administration, increasingly concerned with damping down growing political turbulence, whether of antiwar protestors or of militant poverty area residents. Ultimately there occurred a clear recognition by program decision makers of the connection between administrative change and political change, and in particular, of the risks posed by militant Community Action to the stability of political establishments in major central cities. As a consequence, legitimacy was subsequently withdrawn from the Community Action Program as an instrument for local bureaucratic reform, and a new program known as Model Cities was specifically created to carry out that function. By the close of the Johnson administration in late 1968, CAP essentially retained only the basic function of managing delivery of local antipoverty services, by then mainly prepackaged national emphasis programs.

As to the Model Cities program, its major objective essentially was to restore the urban political executive (mayor or city manager) to a central position in program decision making in order to foster executive-centered reform of the local bureaucracy, rather than reform based on local community organization and action whose form or intensity could be difficult to predict or control. If city public agencies were to be reformed, it now would be a top-down operation ultimately managed by local political elites, and it would be those elites who received the program subsidies and legal authority from federal administrators seeking these reforms on behalf of the poor, rather than the local organizations of the poor themselves. Model Cities will be examined in detail, following a brief discussion next of aspects of Creative Federalism that are important as a context for this sequel to Community Action.

Creative Federalism and Transition in Intergovernmental Relations

Looking back from the perspective of the 1990s, Creative Federalism can be seen as a significant, though brief, transitional stage in the development of postwar intergovernmental relations. Until the mid-1960s Cooperative Federalism was still essentially a system of federal grants administered by the states; the major exceptions were such programs as public housing and urban renewal. Although by 1962 there were 160 federal-state grants amounting to almost $8 billion in funding, direct payments to local governments amounted to less than one tenth of that, about $700 million (Howitt, 1984, p. 7). It was not until the mid- and late 1960s that the flow of federal grant funds relevant to cities expanded dramatically with important consequences for the grant-in-aid system as a whole and for federal-local grants in particular.

Beginning in the mid-1960s the intergovernmental grant system grew rapidly both in number of grants and in amount of federal funding. Congress enacted major new grant programs ranging from Medicaid and aid to elementary and secondary education to urban-relevant grants such as CAP and the Model Cities program, as well as scores of smaller grant programs. During these years new grant programs were established, with federal aid to states and localities increasing from just under $8 billion in fiscal year 1962 to $13 billion in fiscal year 1966. At the end of 1966 there were 379 grant-in-aid programs for state and local government, compared with 31 in 1938 and 160 in 1962. By 1975 some 448 separate grant programs were in operation, distributing approximately $50 billion that year (Howitt, 1984, pp. 5-13).

The change in the grant system was not merely quantitative. By the late 1930s Cooperative Federalism had conclusively removed the barriers to functional interaction between national and state governments by providing for federal funding of a variety of programs administered by state agencies. Emerging in the 1960s as a brief but distinctive subphase of intergovernmental system development, Creative Federalism can be viewed as an outgrowth of Cooperative Federalism, in which the national government now tended to bypass both state and local governments, particularly in order to deal directly with problematic urban areas. As an extension of Cooperative Federalism, Creative Federalism reached down below the level of local government to semipublic organizations such as the CAAs and to neighborhood organizations and private business firms, allowing them to act as agents for the administration of federal grant programs. However, with the expansion of the number and types of categorical programs under the Great Society and of federally delegated authorities for their administration (Sundquist & Davis, 1969), various problems developed for local and state governments, including "fragmentation of effort; lack of coordination among programs; . . . and confusion, delays and red tape" (U.S. General Accounting Office, 1975, p. 9).

As noted earlier, Model Cities was intended as an explicitly urban sequel to the Community Action Program. A key objective was to avoid from the outset the relatively loose coupling that had existed between the local and federal governmental levels in CAP, in which official federal agencies went around the mayor's office to deal with independent local voluntary organizations and paragovernments such as the CAAs. In addition, Model Cities would make a considerably more ambitious attempt at comprehensive planning than Community Action by including physical development planning as well as social services coordination in poverty areas, but this time in a more tightly coupled federal-local official governmental framework. The intention was that Model Cities was to be a second, better-planned, and coordinated attack on poverty in the cities, a program that through improved administrative structuring and procedures would succeed in making Cooperative Federalism work (Sundquist & Davis, 1969).

THE MODEL CITIES PROGRAM:
REORGANIZING THE ATTACK ON URBAN POVERTY

The Model Cities program can be understood best against the backdrop of the two major urban-relevant programs that preceded it—urban renewal and CAP. By the late 1950s urban renewal had come to signify a heavy-handed type of "bulldozer clearance" (Anderson, 1964), tearing down whole blocks of older residential and commercial structures and with them the core of many inner-city low-income neighborhoods. In the process, local resistance to this type of redevelopment was stimulated, leading to an upswing of community organization in low-income neighborhoods, particularly in minority areas. At this stage, urban community organization was frequently a base for saying "no" on behalf of such neighborhoods to further destruction of their physical and social fabric.

By the mid-1960s the organizations of a growing community movement (Bell & Held, 1969) had broadened their horizons beyond this earlier defensive outlook and, with the help of federal antipoverty grants and official recognition, had begun to turn to the larger issue of what they wanted to accomplish positively in the local environment. In the first flush of enthusiasm, local agencies of the Community Action Program proceeded to draw up wish lists of desired changes in community services and institutions, often without much concern for their order of priority or the strategy for their implementation. As a consequence, the lack of either a comprehensive planning vision or practical planning experience added to later obstacles encountered by CAAs as they attempted to bring about change in the organizational domains of local social service agencies. Model Cities administrators were determined not to repeat this mistake.

Program Objectives and Funding

The Model Cities program can be thought of as embodying a new approach to redevelopment in urban poverty areas, superseding both urban renewal and CAP. Through Model Cities, there was to be for the first time explicit government support for comprehensive, multidimensional urban neighborhood redevelopment. This would combine aspects of the social program approach first explored by CAP and a scaled-down type of physical renewal that emphasized rehabilitation rather than massive clearance in distressed residential areas. In short, the new program would be an experiment in sensitively redeveloping the total neighborhood environment, social as well as physical. Overall, it was to serve as a model for the planning and coordination of interrelated programs, concentrating its resources on a relatively few major cities. Reflecting the negative official reactions to the CAP experience, there would be a deemphasis of citizen participation and a return to the theme of service coordination through local Model Cities agencies, together with a new objective of comprehensive planning based on specific program guidelines (Gilbert & Specht, 1977).

Model Cities was initiated officially in November 1966 when President Lyndon Johnson signed the Demonstration Cities and Metropolitan Development Act (later to be known as the Model Cities Act). This occurred in the midst of a period of militant challenge to central authority in cities, sometimes conducted in the streets (often aimed at the mayor's office and city agencies) and frequently engaged in by resident organizations acting in the name of the Community Action Program. However, this was a time of turmoil not only in the streets but also in the interorganizational fields of the cities, affecting both social service systems and municipal government institutions. In these regards, Model Cities could be viewed as a possible new start, aimed at combining more sensitive physical renewal with expanded social services and improved service coordination while avoiding the turbulence of the Community Action Program.

The opportunity for a fresh initiative in urban policy making was a rare one. The Department of Housing and Urban Development had been established just in 1965; later that year President Johnson called together a task force under the chairmanship of Professor Robert Wood for advice regarding organizational structure and programs for the new agency. In the circumstances, an innovative program that addressed important problems of urban poverty could help win the department visibility and a central identifying mission. Optimistically viewed, with the help of strong presidential support a well-designed program might serve as an important step toward rationalizing and coordinating the national system of urban policy making while gaining the needed support of diverse political and bureaucratic interests (Frieden & Kaplan, 1975).

At the time, President Johnson's preference was for the creation of a major new housing program to be incorporated in the further redevelopment of city slums, an objective that had been neglected seriously by the existing renewal program. The Wood task force broadened this objective and recommended a comprehensive program of social as well as physical redevelopment in the slum areas of a limited number of demonstration cities. The report of an earlier study panel in 1964 had transmitted a recommendation for massively concentrating resources on no more than three cities. However, the 1965 task force believed this was too small a number to attract sufficient support for congressional approval. As a result its initial report recommended a total of 36 cities, which increased to 66 in the final report; largely due to trading of favors in the congressional legislative process, that number finally more than doubled (Frieden & Kaplan, 1975, pp. 38, 49, 215).[8]

Such an expansion of program targets was not unusual; in the case of CAP the initial intention had been to concentrate on approximately a dozen demonstration communities, but the program that Congress passed included about a thousand. However, although expansion beyond a limited number of pilot projects was not unprecedented, it is questionable that this was helpful to achieving major program objectives. I will consider later what effect such dispersion of resources had on the Model Cities program.

Model Cities began with the explicit premise that program action had to be preceded by comprehensive, coordinated planning. All cities were invited by HUD to apply competitively for planning grants. By spring of 1968 approximately 350 cities had applied; by the end of 1968, 150 cities had been selected to receive planning grants on the basis of the quality of their applications. Each winning city then would be expected to engage for approximately a year in a complex three-part process to produce a *comprehensive demonstration plan* (CDP). (Half of these cities—the first of two rounds of 75 cities each— already had been selected and begun planning by early 1968.) The CDP required first, a full analysis of local problems and strategies for their solution; second, projections of 5-year action objectives and costs; and third, a detailed statement of first action year costs and administrative arrangements (Gilbert, 1974, p. 182). Approval of this plan by HUD would lead to granting of program operating funds in the form of annual action grants to a *city demonstration agency* (CDA) over a 5-year implementation period. In sharp contrast to the Community Action Program that had provided for no advance planning, this clearly proclaimed the intention of the Model Cities program to operate on the basis of long-term planning and related implementation for any participating city.

Model neighborhood areas designated by the selected cities would constitute demonstration sites for the comprehensive renewal of entire urban slum neighborhoods. Funding support for the program was to be provided through a different type of federal grant than the relatively narrow-purpose categorical grants usually employed for urban renewal and social service programs. Presumably, Model Cities would have priority access to all existing categorically funded federal programs relevant to its projects in participating cities. In addition, there would be a special funding arrangement through which Model Cities also would receive supplemental federal funds in the form of a block grant that "could be spent flexibly and unhindered by normal federal grant-in-aid restrictions" (Frieden & Kaplan, 1975, p. 38)[9] in accordance with locally established program priorities. These supplemental funds would be used to help the city pay 80% of its required matching share to receive existing federal categorical grants relevant to local program objectives or for new locally designed program activities. The intention was that this would allow for a more appropriate mix of projects tailored to specific community needs and encourage greater innovation in local program planning. This arrangement was not unlike the use of the Community Action block grant to develop local initiatives programs or pay for federally developed national emphasis programs. In both cases, Creative Federalism used block-type grants to access more flexibly the existing categorical system or to pay for development of innovative local projects.

With comprehensive objectives and flexible funding, Model Cities promised to be a key program in a revised approach to urban redevelopment. As a major initiative of the new HUD agency, it opened up the prospect for a Department of Housing and Urban Development that would initiate long overdue improvements in several respects: greater emphasis on the often neglected needs of inner-city residential areas following more than a decade of emphasis on CBD economic revitalization; a more sensitive approach to residential redevelopment that emphasized rehabilitation, rather than the often neighborhood-damaging bulldozer clearance strategy; and a new responsiveness to needs for social as well as physical renewal within targeted residential renewal areas. However, the full realization of these new prospects would depend significantly on the introduction of improvements in program planning and administration, with particular attention to the federal-local relationships involved.

Program Administration

In regard to program administration, each participating city would be required to designate a city demonstration agency with official authority to implement in a coordinated manner all the phases of the program. In order to avoid repeating the political conflicts of the Community Action Program, the decision was made that the chief public executive (mayor or city manager) would have the power to review and sign off on the project applications of the CDA before they went on to HUD in Washington. Organizationally, this meant that the Model Cities program would be conducted as an intergovernmental relationship in which program resources would flow between established public agencies at the federal and local levels, rather than (as in the Community Action Program) between the federal granting agency and community organizations over which local public officials often had only loose control. In sharp contrast with OEO's community action agencies, the Model Cities program's local administrative machinery was required to be a public agency (the CDA) under the formal authority of the mayor's office, with the mayor appointing the CDA director.[10]

These provisions responded to concerns for tightening up program management in a federal system generally characterized by rather loose coupling between the federal and local levels of decision making and hence a low degree of coherence in policy implementation (Robertson & Judd, 1989). This included some reduction of the emphasis on citizen participation that had been a major focus of the Community Action Program; however, just how much citizen participation would be appropriate to this new program framework was not immediately clear. Initially, HUD guidelines for Model Cities in 1966 made no mention of power sharing with local residents; rather they referred only to creating the means for "communication and meaningful dialogue" between residents and the CDA. This suggested reducing citizen participation to an essentially advisory role for residents, through which they could provide input to the CDA decision-making process without any real power of decision over its planning products. Instead of endorsing maximum feasible participation, the legislation that authorized Model Cities referred ambiguously to "widespread" citizen participation, and the 1966 Program Guide observed vaguely that CDAs should provide for area residents to have a "meaningful role" in local program policy making (U.S. DHUD, 1966, pp. 11-14).

In part, these ambiguities reflected a division between those officials associated with the program who favored a sharing of power between city hall and neighborhood residents and those who felt residents should have no more than an advisory role on the CDA policy board (Gilbert & Specht, 1977). In either case, after the experience of the Community Action Program there could be no question of simply omitting citizen participation; it was now necessary to at least include a participation component in order to win the support of residents whose neighborhoods were targeted for redevelopment.

Clearly, the resident participation movement enlivened by CAP was still vital; its legacy had taken root in numerous cities by the beginning of the Model Cities program.

At the same time, it was clear also that the major purpose envisioned for Model Cities was that it serve as an improved delivery system for efficiently channeling existing federal programs to participating city agencies (Ripley, 1972). HUD's emphasis was on developing a chain of program funding and delivery linkages between official institutions of government, not between federal agencies and local neighborhood organizations. Although citizen participation would be encouraged, it would not be supported by HUD as a prime objective of the program, and there would be no further support for maximum citizen participation in the sense of resident organizations exercising major control over local program policy.

Program Types and the Local Organizational Field

In any event, the precise degree of resident influence in the CDA was not something that could be determined unilaterally by HUD's Model Cities Administration in Washington. Several patterns of influence regarding the CDA planning system emerged; which one applied to any given CDA was significantly affected by the local interorganizational context set within the framework of HUD program guidelines and preferences regarding resident participation. The major actors in this context were the mayor, the CDA director and planning staff (all accountable to the mayor), and the resident organizations and any related technical staff they may have had available to them. At stake was the degree of control over the CDA's neighborhood renewal planning process exercised by city hall as against the resident organizations.

The resulting structure of the local Model Cities program can be understood by recognizing that the CDA played the role of an intermediary planning organization standing between its major clients, city hall on the one side and the neighborhood resident organizations on the other. CDAs whose staff identified with city hall, either strongly or moderately, would tend to treat the chief executive as their primary client. Hence such CDAs were classified as mayorally oriented *staff-dominant* or *staff-influence* CDA types; residents had relatively little influence in those planning systems. On the other hand, CDAs that identified with the neighborhood resident organizations, either strongly or moderately, could be thought of as *resident-dominant* or *resident-influence* types. Alternatively, the CDA could engage in a balanced relationship known as *parity*, not identifying with either client more than the other or acting as the planning agent of either one but as a relatively impartial bargaining arena and mechanism for planning. This was the CDA planning system preferred by HUD planners in Washington.

Despite initial concerns regarding a repeat of the widespread turbulence that had accompanied citizen participation under the Community Action Program, staff-dominant/influence CDAs (characterized by relatively weakly organized, passive resident participation and by CDA staff orientation to the mayor's planning preferences) made up a majority (56%) of CDAs in the 142 cities for which HUD ratings were obtained by the end of two nominally yearlong planning rounds. This was evidence that unlike CAP the Model Cities program was in good part a "mayor's program," in which the perspective of the city's chief executive was translated into local program policy through the appointed CDA director and technical staff. At the same time, although the parity CDA relationship (associated with coherent, politically integrated residents' organizations) may have been preferred by federal program planners in HUD's Model Cities Administration, it accounted for only about one fourth (26%) of all CDAs. Finally, resident-dominant CDAs (associated with well-organized, strongly militant resident groups generating high levels of local intergroup conflict or "environmental turbulence") accounted for less than one fifth (18%) of CDAs (Gilbert & Specht, 1977, pp. 66-67).

This testifies to the significance of the local interorganizational context for shaping the CDA planning system particularly during the initial, planning stage of the Model Cities program. Later program data suggest that HUD was beginning to succeed in shifting unstable influence types toward the parity model (U.S. DHUD, 1973). However, the change in presidential administration in January 1969 after only two years of program operation effectively removed the opportunity to further support such a trend.

Model Cities in the Transition From the War on Poverty

During the year after coming into office in January 1969, the Nixon administration began a reorganization of the Model Cities program that involved objectives that ultimately went far beyond the program on behalf of goals and constituencies sharply different from those targeted by the War on Poverty. These objectives focused on changing the ground rules for administering antipoverty programs inherited from the preceding administration and on developing new intergovernmental arrangements that would serve as the basis for a New Federalism—one that would include major changes in the orientation of urban policy with important implications for issues of urban poverty.

Significantly, the first steps in this direction were associated with Nixon administration moves to end decisively any ambiguity about citizen participation. New guidelines issued in December 1969 gave final responsibility to the "chief executive and the local governing body" for the "development, implementation, and performance of the Model Cities Program" (Gilbert & Specht, 1977, p. 17; U.S. DHUD, 1969). Resident organizations were excluded from this formulation, which effectively left them with at most an advisory role. Given the usually close ties of CDA staff to mayor, this would tend to make the staff-dominant type the prescribed pattern of influence in CDAs. Shortly thereafter, another guideline broadened the principle of local representation well beyond the model neighborhood to include "representatives of major elements of the city as a whole (e.g., religious, charitable, private, and public organizations) having an interest in the Model Neighborhood" (Gilbert & Specht, 1977, p. 17; U.S. DHUD, 1970).

In general, the effect was to dilute residents' influence on CDA planning for their neighborhood by introducing a wide variety of outside interests into the planning process, regardless of whether those interests (ranging from religious organizations to real estate developers) shared the concerns of the affected area's residents. Together these guidelines made it clear that the thrust of Model Cities had changed from a program initially encouraging local resident participation and even a degree of control to one that would tend to minimize resident influence in planning the neighborhood's redevelopment.

In the next phase of program reorganization, known as *Planned Variations* (first announced in September 1970), the Nixon administration took an important further step in the reshaping of Model Cities by authorizing a series of major variations from established program practices. These changes were instituted in a sample of 20 Model Cities in July 1971. They included, notably, the elimination of most HUD reviews of the local program (except those written directly into law by Congress). This had the effect of significantly decentralizing program administration, largely shifting it from the federal to the local level. In addition, they gave the city's chief executive the right of review and sign-off on all federal categorical programs applied for by city agencies, "prior to their approval by federal agencies" (Gilbert & Specht, 1977, p. 20).

Together these changes meant that authority for federal program application and oversight now clearly rested with the mayor or city manager, rather than the city's separate public agencies, whose relative independence as "bureaucratic fiefdoms" long had been nurtured by categorically funded federal grants. Presumably, this offered the opportunity for improved coordination between federally funded programs under the oversight of the locality's chief executive. In addition, in the approximately two dozen cities designated for Planned Variations, the Model Cities program was permitted now to include the entire city as its planning area, rather than targeting specific disadvantaged neighborhoods. That variation meant the deep reduction of poverty neighborhood influence on the CDA planning system, and it removed any remaining possibility of neighborhood-based challenge to mayoral control of the Model Cities program, on behalf of antipovery priorities.

Viewed narrowly, such measures simply converted the designated Planned Variations cities into experimental areas for a locally administered, city hall-controlled program. But in fact they did more than this. Overall, they provided federal authorization for Planned Variations to serve as a pilot program within the larger Model Cities program, testing a basically new approach to policy planning and implementation—one that would be applicable not only in the area of urban policy but in a broad range of intergovernmental policy. Planned Variations was designed to take the Model Cities program—a relatively advanced product of Creative Federalism, structured along the lines of block grant—and reorganize its administration to fit the objectives of a projected new system of federalism. In specific, Planned Variations was created to test the capacity of local government for discharging new responsibilities that it would have to carry under a significantly decentralized, executive-managed federal system. Of particular importance, the Planned Variations experiment would help evaluate the feasibility of a critical trade-off: acceptance by local government executives, city agency leaders, and their respective staffs of new roles and expanded workloads in grant administration and coordination in return for the promise of increased access to federal program resources and the opportunity to apply them flexibly with relative independence from federal oversight.

Federal policy decentralization of the type suggested by the Planned Variations experiment was at the heart of President Nixon's plans for a New Federalism. Its ultimate objectives were to reduce significantly the influence of the categorical grant system (whose expansion owed mainly to the liberal program initiatives of the New Deal and Great Society) and to reduce the power of federal regulation in the implementation of intergovernmental policy. Multiple functionally related categorical grants were to be consolidated into block grants in broad functional areas such as community development, manpower, health, education, and social services. This in turn would reduce the numbers and influence of categorical grants, which since the War on Poverty increasingly had become targeted to (and encouraging to the organization of) previously unorganized or disadvantaged constituencies such as the urban poor.

The Model Cities program was chosen as the framework for Planned Variations of the type discussed because, even more than Community Action, it had already adopted the use of block grant type federal-local funding arrangements to improve access to existing categorical grants and to encourage local flexibility and innovation in the use of federal funds for combating urban poverty and blight. But Planned Variations went several steps further to remove the power of federal review over local program operation, which had been retained specifically under Creative Federalism programs in order to assure that those programs remained targeted to using their funding to improve the condition of their intended recipients.

Now the block grant precedent of Model Cities would be used as a stepping stone to a relatively open-ended sharing of revenue (special revenue sharing within broad block categories such as education or community development) between the federal and local levels, with little focus on whether those federal revenues were spent for the full benefit of poverty areas and their residents. In addition, almost wholly unrestricted general grants under general revenue sharing would contribute to reducing further the regulatory powers of federal agencies over city and state governments receiving federal funding. At the time, those objectives were justified in terms of offering opportunities for improved grant planning and coordination; but to be understood fully they also must be examined closely in terms of their implications for the processes of urban social-economic and political development and for whose ultimate benefit those processes were to be oriented. I will continue with that analysis in the next chapter.

NOTES

1. According to Friedman (1967), writing of the War on Poverty: "Clearly the war would not be fought with transfer payments alone," such as the Social Security payments for income maintenance of the elderly introduced earlier by the New Deal. There would be "no massive redistribution." Contrary to images promoted by its opponents, "the poverty program itself . . . did not call for major spending. . . . The Economic Opportunity Act of 1964, according to Vice-President Humphrey, writing in 1966, 'modified' the 'older dominance of transfer payments,' with an 'increased emphasis on job preparation' " (p. 36; Humphrey, 1966, p. 12).

Friedman (1967) noted that the hope of the antipoverty planners was that "new techniques [of programming for expanded economic opportunity] would supplement and ultimately replace 'transfer payments'; otherwise the war on poverty was meaningless." As he further observed, "the Economic Opportunity Act (EOA) by-passed the aged poor" (p. 37), who depend on transfer payments to maintain their incomes and, according to Levitan (1969), "could not be motivated, trained, or 're-habilitated,' " hence EOA "concentrated on helping young people" (p. 11).

2. There has been considerable debate concerning the sources of the War on Poverty. As Wilson (1987) observed, conventional accounts have focused on the role of social critics such as Galbraith (1958) and Harrington (1962) in bringing the facts about the prevalence and severity of poverty to light in the late 1950s and early 1960s, or they have emphasized the importance of revelations such as those concerning White poverty in Appalachia to President Kennedy's plans for his 1964 legislative program (Sundquist, 1969). On the other hand, some observers have pointed to the significance of racial issues and the civil rights movement in placing the issue of poverty on the federal agenda. See Wilson (1987) for further discussion and bibliography.

3. For a combination of political and economic reasons, President Johnson became committed strongly to providing the tax cut first proposed by Kennedy. Even after it "was enacted and as [Secretary of Labor] Wirtz had insisted would be the case, it failed to wipe out unemployment," Johnson totally rejected proposals for public job creation, labeling it as little more than "make work." As a consequence, noted Sundquist (1969), "the War on Poverty remained harnessed exclusively to the 'services strategy' or 'opportunity strategy' " (p. 244). Regarding the debate within the Johnson administration about the need for a direct public employment program in fighting poverty, see Katz (1989).

4. Citizen participation is characterized by ambiguous possibilities. As Levi (1974) observed, "Social programs are a forum for action as well as a program for social control. . . . [T]he state, i.e. government and government-supported research, services and agencies, gives rise to and limits poor peoples organizations, and . . . [those] organizations may in turn elicit the social involvement of the state" (p. 78), by exerting pressure on it to act on their behalf.

5. Such coordination was inherently a difficult task, taking into account the different programs available to the target area from both the local and federal levels. Locally, a city might get services from a variety of different functional agencies associated with different jurisdictions; for example, "a county public health department, a state department of public assistance, a state agency for unemployment compensation, and a local board of education" (Gilbert & Specht, 1977, p. 5). Coming from the federal level, programs managed by OEO included the Job Corps (a residential job training program for teenage youth) and Vista (Volunteers in Service to America, the domestic equivalent to the Peace Corps), as well as a series of national emphasis programs widely applicable to poverty areas nationwide, such as Head Start, Upward Bound, Legal Services, and the Comprehensive Health Service. At the same time, the Department of Health, Education, and Welfare operated the Work Experience Program (a training and employment program for welfare recipients and others) and the College Work-Study Program, while the Department of Labor operated the Neighborhood Youth Corps, an at-home program of training and job experience for local youth (Plotnick & Skidmore, 1975).

6. As Henig (1985) noted, "Although the War on Poverty was characterized as representing an overly centralized federal intrusion into state and local affairs, *the major mechanisms for implementing the programs were highly decentralized*" (p. 104). Thus, in programs such as Title I of the Elementary and Secondary Education Act, the states had broad discretion regarding the application of federal funds; at the local level, city governments and school districts had much leeway in administering a variety of educational and training programs. In fact, decentralization was carried even further in other antipoverty programs such as the Community Action Program, which bypassed local officials and placed a substantial degree of administrative discretion in the hands of community action agencies and their participating community organizations in target areas.

See also Miringoff and Opdycke (1986). In concluding their careful discussion of the antipoverty aspects of the Great Society, they observed that "the important central role played by the federal government in initiating and pushing through new programs was offset to a considerable extent by the decentralizing effect of dealing directly with thousands of small organizations, neighborhoods, and interest groups" (p. 64).

7. As products of Creative Federalism, the Community Action Program and the subsequent Model Cities program were structurally transitional programs that incorporated categorical grant features associated with Cooperative Federalism and block grant features associated with New Federalism. According to the Advisory Commission on Intergovernmental Relations (1976), "In their broad functional scope, target grants bear considerable similarity to block grants. . . . Yet, their other traits—especially the "targeting" of assistance and stringent Federal conditions—resemble more the characteristics of some categoricals" (p. 86).

8. As James (1972) observed, "The base of support certainly was broadened by inclusion of most major cities ([including] all but five of those over 300,000) and 45 states, along with several 'deserving' smaller towns in the congressional districts of committee chairmen" (p. 74).

9. The block grant feature would appear in two forms. The "full array of existing grants under all relevant federal programs" would be made available to the program on a priority basis. To increase the recipient city's flexibility in implementing specific local objectives, these grant funds would be merged in a common account, to be applied at the discretion of the local CDA, much like a block grant (Frieden & Kaplan, 1975, pp. 45-46).

10. Despite the formal power of the mayor to appoint the CDA director, this still left the director and CDA staff to cope with often competing demands from city hall and the target neighborhood organizations. Particularly where neighborhood groups were well organized and militant, this competition was not always wholly resolved in the mayor's favor (Sundquist & Davis, 1969, pp. 94-99).

[7]

New Federalism and the
Reorganization of Urban Policy

THE NIXON ADMINISTRATION
AND THE NEW FEDERALISM

On its most basic level, President Nixon's New Federalism might be described as a concentrated effort to dismantle or diffuse the programs associated with the War on Poverty and curb whatever influence they retained over the objectives and organization of the federal policy system when he entered office in 1969. As evidence of this effort, from its beginning the Nixon administration directed itself to the organizational nerve center of the War on Poverty, abolishing the Office of Economic Opportunity and neutralizing the programs that had been associated with it.

As I have noted, programs such as Community Action provoked intense interorganizational conflict when they challenged the organizational domains of municipal agencies in order to improve the administration of services to the poor. Now they were relieved of their earlier mandate to organize low-income residents and confined essentially to the role of routine provision of services, often with deep cuts in funding. Or, as in the case of Model Cities, they were used as experimental proving grounds for converting antipoverty efforts to New Federalism initiatives. Programs such as Head Start, which had received broad support as channels for providing new social services, were not abolished but were fielded out to established departments. Others such as the Legal Services Program, which became controversial through its assistance and advocacy on behalf of the poor, experienced both removal from OEO and reduction in approved activities and funding.

In his second term, following a landslide reelection, Nixon adopted an even bolder strategy—the impoundment of funds that he refused to spend for programs of which he disapproved, including programs of major relevance to city development such as urban renewal, Model Cities, and public housing. Although the Nixon administration later released these funds under court order, the construction of federally subsidized housing, for example, was reduced significantly for several years. Nonetheless, Nixon did not merely conduct a negative campaign against the programs developed by the War on Poverty. As an astute analyst of intergovernmental relations, he recognized how important the organization of the federal system is to enabling or constraining the implementation of public policy objectives, especially those concerned with economic redistribution and related social change (and most particularly those bearing on highly conflictual social-economic issues, such as race relations). The antipoverty programs of the Great Society had represented a major contemporary adaptation of Cooperative Federalism to liberal social and political objectives. In removing those programs, Nixon saw a unique opportunity to put in their place the administrative tools that would enable the reshaping of the federal system and the redefinition of future options for urban policy.

Each major domestic initiative that Nixon put forward in place of the War on Poverty was associated with reorganizing the federal system so that its very structure, embodied in the pattern of intergovernmental program relationships, would facilitate an orientation away from redistributive objectives and their supporting constituencies (including various urban officials as well as the urban poor) and toward more conservative constituencies who mainly were not urban, and mostly not poor. At the same time Nixon's new domestic initiatives would provide compensatory subventions, if not to the poor, at least to their local officials. These initiatives found their clearest expression in the Nixon administration proposals for a New Federalism.

REVENUE SHARING:
GENERAL AND SPECIAL

The New Federalism was conceived essentially in terms of two types of revenue sharing. One was *general revenue sharing* (GRS), designed to distribute federal revenues to state and local governments with virtually no binding restrictions as to how they were to be spent. This was in sharp contrast to categorical grants, which carry specific conditions regarding their use in order to assure that they reach their designated targets. GRS monies could be used to fund any regular local government activity, or they could be used simply for tax relief if so desired (Frieden & Kaplan, 1975).

The other was *special revenue sharing* (SRS), initially aimed at consolidating over 100 existing categorical grant programs into 6 large block grants in specified functional areas such as education or community development. If successful, later consolidations would follow in a process intended to shrink the categorical grant sector that had been used extensively during the War on Poverty to ensure the targeting of funds to poverty areas, particularly in large cities.

General and special revenue sharing were proposed as complementary parts of the scheme for a New Federalism presented by President Nixon in his State of the Union message in 1971. With regard to intergovernmental relations, this was an extremely important proposal, aiming at a basic shift of power relations in the federal system through fundamental changes in the organization and funding of intergovernmental programs.

In the first year $16.1 billion would be distributed; $5 billion through general revenue sharing and $11.1 billion through special revenue sharing in six functional areas. General revenue sharing was viewed widely as "new money" (Haider, 1974, p. 256), an additional infusion of funds into the federal grant system that Nixon initially promised would not come at the cost of reducing existing categorical programs. Special revenue sharing, on the other hand, was seen as an encroachment on the categorical grant system, which since the New Deal had enabled the establishment of a growing number of fiscal linkages between Congress and its various interest-constituencies throughout the nation. As a consequence, members of Congress could be expected to be much less receptive to this type of revenue sharing and quite selective about which aspects of it they might approve, if any.

General Revenue Sharing

First to be adopted was general revenue sharing in 1972 through passage of the State and Local Fiscal Assistance Act. Under its terms a total of $30.2 billion would be provided to state and local governments over a 5-year period for an average annual allocation of $6 billion. One third of each year's allocation would go to the 50 states and the remaining two thirds to the approximately 39,000 general-purpose local governments in the nation. For most cities that had participated in the Model Cities program, the annual revenue sharing funds would equal at least what they had received under Model Cities grants (Palmer, 1984, p. 27). However, the uses to which they would be put would not emphasize necessarily the same priorities as that program.

General revenue sharing was used by the Nixon administration as an important opening wedge to adoption of the overall New Federalism proposal because it was attractive to local officials as federal grant money that imposed no requirements for local matching funds. In effect, it was an automatic entitlement program funded not on the basis of an application but simply by formula, for all general-purpose subnational governments in the country (Dommel et al., 1982). Of particular importance for the passage of GRS was the "growing influence of a new form of interest group residing within government itself: the intergovernmental lobby, composed of associations of mayors, governors and other state and local officials" (Conlan, 1988, pp. 65-66). The intergovernmental lobby, made up of elected officials outside the Congress, played a critical role in the enactment of GRS by working intensively to pass the authorizing legislation. For governors, mayors, and county officials, GRS arrived at a time of growing fiscal stress during the early 1970s. Rising public service costs, coupled with inadequate revenue-raising authority, created unavoidable fiscal pressures on local officials that persuaded many formerly reluctant officials (especially local Democrats) to accept the Nixon proposal for GRS and lobby on its behalf as they had not lobbied when it was first proposed in 1971. GRS provided a welcome supplement to local budgets and a margin of relief from pressures to increase taxes to cover the growing cost of public services (Clark, Iglehart, & Lilley, 1972a, p. 1557). At the same time GRS also involved a shift of control over the spending of federal funds, away from program professionals and related agency specialists to elected officials and other policy generalists. On both these counts it was greeted favorably by state and local politicians, particularly in the increasingly difficult fiscal situation of that period.

Among public officials and related professionals, support for general revenue sharing was broad based, bringing together city-oriented organizations (the U.S. Conference of Mayors, the National League of Cities, and the International City Management Association) and county and state organizations (the National Association of Counties, the Council of State Governments, the National Legislative Conference, and the National Governors' Conference). This intergovernmental lobbying coalition, known as the "big seven," included mayors, city managers, county executives, and governors and represented an impressive array of experience and influence regarding state-local affairs. The opposition to GRS consisted of most of organized labor, represented by the AFL-CIO (with the notable exception of the American Federation of State, County, and Municipal Employees), civil rights groups, and urban liberals. Their major concerns were that such open-ended sharing of federal revenues with states and localities would bypass minority and working-class constituencies because "without specified and enforceable federal performance standards, there is no assurance that federal civil rights guarantees and fair labor practices will be applied to projects supported by no-strings federal grants" (Judd, 1988, pp. 335-336).

Despite strong ties to organized labor, with many cities experiencing increasing fiscal stress by the early 1970s, Democratic party leaders—aware of the importance of the party's local electoral base—moved to advance a legislative alternative to the Nixon administration's revenue sharing proposal. With the aid of Wilbur Mills, the Democratic chairman of the House Ways and Means Committee, they struck a compromise with the Nixon administration that gave local governments two thirds of the funds. This reversed the relative state-local shares of program funding originally proposed by the administration and won the support of younger Democratic members of Congress who were less reliant than senior Democrats on categorical grant linkages to their constituents and "more vulnerable to the high-pressure lobbying by mayors and governors" connected with the influential intergovernmental lobby (Conlan, 1988, pp. 68-69, 71).

GRS funding for cities was determined by a formula that took into account such factors as population size, per capita income, and tax effort. For a city to do especially well it had to be large, have a relatively large percentage of its population below the current poverty line, and have a high tax effort relative to its economic base. Other things being equal, those communities with comparatively high local tax rates would be rewarded with larger grant allotments by GRS. Funding for states was distributed on the basis of two formulas similar to that for cities, each again including population, per capita income, and tax-effort factors, although in somewhat different combinations. The formula constructed in the House of Representatives tended to favor "urbanized states with high-tax efforts and large concentrations of poor, such as California and New York," while the Senate formula favored "rural states" with "large concentrations of poor," such as those in the South (Clark et al., 1972a, p. 1556). By congressional compromise, each state's revenue sharing funding would be based on the formula that allotted it the most money.

Although a key factor in each of these formulas was per capita income (and related to it, the incidence of poverty within the given jurisdiction), that affected only the amount of revenue sharing money that was received locally but not necessarily how it would be spent. Typically, local governments spent most of the funds they received on functions other than social services. Large cities (over 250,000)—many of which had operated substantial antipoverty service programs during the 1960s—now put over two thirds of GRS funds into expenditures directed solely to the basic services of public safety (police and fire services) and transportation. Among all cities, most put no more than 1% of their general revenue sharing funds in social services; among large cities social service spending from GRS funds was not much greater, only 2% (Judd, 1988, p. 342). These spending patterns were the first suggestions of a clear shift away from the priorities of the War on Poverty.

Further, although it was described repeatedly by the Nixon administration as a way of bringing power (and presumably, supporting financial resources) closer to the people, GRS must be understood as an important shift from redistributive to distributive fiscal policy for states and localities. It was described appropriately as a type of distributive localism (Beer, 1976), through which every subnational general-purpose government received some funding. A key by-product of this type of universal entitlement program was the spreading and thinning out of federal funds that might otherwise have been concentrated on areas of greatest fiscal need.

In the GRS program, economically depressed big cities found themselves lumped together as recipients of program funds with numerous affluent suburbs in much better fiscal condition. In addition, the use of population size in the GRS formula meant that many of those suburbs would be rewarded further each year because their populations were growing, while older cities repeatedly would be losers because their populations were in rapid decline. As a Brookings Institution study observed: "Problems of the most troubled central cities are not in any major way ameliorated by general revenue sharing" (Nathan & Adams, 1977, p. 107). Such amelioration was indeed not the objective of GRS, which was structured instead to direct resources toward communities enjoying growth, rather than toward communities in need and typically losing population.

Special Revenue Sharing and Block Grants

The second major component of the New Federalism was represented by Nixon's proposals for special revenue sharing in the functional areas of education, job training, urban community development, rural development, law enforcement, and transportation. Within each area, SRS essentially consolidated previously existing categorical grant funding and drastically reduced

outside controls over local spending discretion. In the community development grant for example, funds for urban renewal, Model Cities, and about a
half dozen other urban improvement grants were combined in a single block.
So long as it acted consistently with the block grant's overall objective, the
recipient community could use grant funds more or less as it saw fit. Community planners now could decide themselves what relative emphasis to place on
the various specific purposes covered by the grant, rather than being bound
to spend according to the priorities previously spelled out categorically. This
encouraged the support of many local officials who felt it freed them to plan
according to their own ranking of local needs.

Together the six block grants were to compose a package of revenue sharing
measures that would provide increased "flexibility in spending federal grants
within broad functional categories" (Frieden & Kaplan, 1975, p. 239). As
proposed by the administration in 1971, SRS funds would be determined by
a formula incorporating measures of population size, extent of poverty, and
overcrowded housing as an automatic entitlement to all eligible governments.
No application would be required; hence there would be no need for review
and approval by the federal granting agency. Had this passed as proposed,
special revenue sharing would have closely resembled general revenue sharing
in providing "a way of transferring funds from the national to the local level
without further federal decision making" (Reagan & Sanzone, 1981, p. 128),
an idea that made it quite attractive to local officials.

However, Congress found itself unable to go along with this concept for a
number of reasons. First, it was reluctant to give up its right to review the
details of how revenues to be raised for broadly designated purposes under
special revenue sharing actually were spent. Although it agreed that spending
discretion of governments receiving SRS funds was to be considerably greater
than that for categorical grants, Congress did not wish to pass SRS simply as
an extension of open-ended general revenue sharing. Furthermore, Congress
did not rush to support Nixon's proposals for special revenue sharing because
of concerns that SRS had the potential of diminishing the relative importance
of categorical grants and with them some important sources of congressional
power. If applied to a wide variety of policy areas, special revenue sharing
would have eliminated many of the mutually supportive connections built up
between Congress and the various constituencies benefiting from its categorical grant programs (Conlan, 1988). Even confined to the areas proposed, SRS
funding by automatic entitlement could set a precedent that ultimately would
weaken congressional oversight of the operations of executive branch agencies
slated to administer the proposed block grants. In the area of urban development,

this would have involved especially HUD and the then Department of Health, Education, and Welfare as key agencies supervising grants of significance to urban communities and their populations.

Because Congress continued to hold back from approving President Nixon's special revenue sharing package, he moved to ease the concerns of opponents by reducing the scope of his proposals, cutting back from six block grants to four (thereby consolidating fewer of the categoricals originally proposed for elimination). As an added concession to Congress, the revised proposals allowed for federal review and approval by HUD of local block grant applications.

Nevertheless, these revisions failed to produce quickly the desired congressional acceptance, whereupon Nixon proceeded in 1973 to impound grant funds already appropriated by Congress for programs such as urban renewal, Model Cities, public housing, and sewer construction, most of which had direct relevance to urban needs. Such actions produced a strongly negative congressional response and reduced the likelihood that even the scaled-down SRS package would be approved. They raised the question whether Nixon's basic objective simply was to organize more rationally the provision of federal grants to states and localities or whether there was also a hidden agenda to cut back on the total funding that would flow through the new grants. In short, was the Nixon proposal essentially intended to improve the management of the federal grant system that funds assistance to urban and other domestic needs, or did it also aim at gradually defunding that system?

Significantly, of the remaining four proposed block grants—community development, job training, education, and law enforcement—the first three covered functional areas that had been central to the War on Poverty. Only the first two of these grants finally were enacted 2 years after the passage of general revenue sharing. One factor that helped them win enactment despite an atmosphere of considerable political distrust was the fact that both the community development and job training proposals had the "independent support" of "policy specialists in and out of Congress" (Conlan, 1988, p. 254).

This tended to support the image of reforming the federal grant system for the sake of better management and was particularly helpful in countering congressional distrust of the administration's political motives. Also distinctive of these two block grant proposals was that they had special significance for urban areas; this was a factor in the strong backing that they received from the intergovernmental lobby, notably the mayoral organizations (the U.S. Conference of Mayors and the National League of Cities). They were passed respectively as the Comprehensive Employment Training Act (CETA) of 1973, and the Community Development Block Grant (CDBG), which was part of the Housing and Community Development Act of 1974 (Conlan, 1988).

Comprehensive Employment and Training Block Grant

The Comprehensive Employment and Training Act of 1973 consolidated 17 categorical grants in the area of job training and development and provided countercyclical funding for public service employment to help combat the unemployment effects of recessionary swings in the business cycle. The act provided money to local governments (cities or counties over 100,000 population) to operate as prime sponsors of a variety of employment-related approaches—from counseling and training to placement and employment—that would fit best the employment needs of the local area (Palmer, 1984, p. 28). State governments could act as prime sponsors for those areas in the state not covered by these other eligible sponsors. Prime sponsors were required to clear their plans with the federal Department of Labor before they received funding, but they were allowed considerable discretion in designing and administering specific programs for the locality (Kettl, 1980). Thus despite earlier administration concessions to federal agency review of local program plans, once the local plan was approved the block grant effectively decentralized implementation to recipient local governments.

Consistent with prevailing assumptions that jobs would be available in the economy for trained employees, Title I of the act initially was intended to provide training and wage subsidies for programs to equip particularly the most disadvantaged unemployed for jobs, mainly in the private sector. However, with a recession developing in 1974 soon after the program was established, fewer private sector job slots were open, and employers often were reluctant to hire the "hard-core" unemployed for the remaining jobs available. Thus, program emphasis shifted increasingly toward public sector employment, with state and local governments using CETA to provide funding for jobs in government agencies, public schools, and hospitals (Henig, 1985).

Given the serious economic problems associated with the recession of the mid-1970s, states and localities found themselves under conditions of considerable fiscal stress. In these circumstances they increasingly used CETA funds to subsidize the continuing employment of workers on the local government payroll who now were exposed to possible unemployment due to fiscal stringency. Thus, a growing volume of CETA funding went to individuals who required no significant upgrading of skills to perform their jobs. In effect, the pool of the current or likely unemployed was "creamed" for those with existing skills or previous qualifying employment, and the low-skilled, hard-core unemployed who had been intended as beneficiaries of the CETA legislation were often passed over. Eventually the job training aspect of CETA came to be overshadowed by local government use of the program to fund existing public service positions, amounting to about 300,000 in 1974 and rising to over 650,000 by 1978 (Palmer, 1984, pp. 33, 37). Thus the original objectives of the act were doubly displaced: away from the most disadvantaged unemployed and away from the original focus on job training, with local CETA programs being used instead to support the funding of regular civil service positions on a long-term basis. Ultimately this raises basic questions regarding the block grant emphasis on decentralized implementation, which at the local government level allowed widespread distortion of key CETA program objectives established by Congress (Parkinson, Foley, & Judd, 1989, pp. 18, 19).

We turn now to discussion of the Community Development Block Grant in some detail because of its significance as the major community development grant program since the mid-1970s. Much of the focus is again on issues of targeting and of decentralized local implementation in relation to CDBG program structure and objectives.

THE COMMUNITY DEVELOPMENT
BLOCK GRANT PROGRAM

The Community Development Block Grant Program consolidated seven former categorical grants—urban renewal, Model Cities, and five grants in the area of physical improvement (neighborhood facilities, grants for water and sewage utilities, open space, urban beautification, historic preservation)—together with two related physical development loan programs (loans for public facilities and for housing rehabilitation). Reflecting these components, the major objectives of CDBG stand in contrast to the Model Cities focus on providing social services such as education, job training, and the like, for the support of "human renewal." Although the Model Cities grant was folded into CDBG, strict limits were placed on spending CDBG funds for social services (a maximum of 20% of the total grant), so that CDBG has functioned predominantly as a consolidation of grants in support of physical improvements rather than human renewal, in the local community (Frieden & Kaplan, 1976).

The first funding period for CDBG covered the 3 years from 1975 through 1977 for a total of $8.4 billion; since then program funding has been periodically renewed, although at reduced levels by the late 1980s. The core of the CDBG grant system has consisted of formula-based funding entitlements going to more than 500 metropolitan cities (over 350 central cities and 150 suburban satellite cities—all larger than 50,000 population—located in metropolitan areas) and to over 70 populous urban counties (population greater than 200,000) typically located in the suburban rings of metropolitan areas (Nathan, Dommel, Liebschutz, Morris, & Associates, 1977, p. 114).

Beyond that, nonentitlement communities (under 50,000 population) could apply to HUD for discretionary grants, and a growing amount of CDBG funds went on this basis to numerous small cities in suburban and nonmetropolitan areas, among which many never before had received urban aid. The result was a considerable expansion in the number of communities receiving urban aid nationwide. Less than 150 communities had received funding under Model Cities, and approximately 1,200 communities received some form of urban renewal assistance by 1974. By comparison, in just the first year of CDBG, almost 3000 grants were made. Overall then, CDBG tended to spread and diffuse urban aid considerably more widely than preceding urban programs, particularly to suburban and nonmetropolitan communities (Dommel et al., 1982, pp. 38-39). As two close observers of the program have noted, at its inception CDBG tended to "approximate a general revenue sharing program for public works" (Frieden & Kaplan, 1976, p. 45) and, at the same time, for physical improvements related particularly to housing rehabilitation (U.S. DHUD, 1992, p. A-1).

CDBG at the Local Level

Local Perceptions of CDBG

In general, block grants tend to allow localities easier access to grant funds and broader discretion in using them, compared with categorical grants. CDBG has been no different in these respects. Hence, from the perspective of local officials, CDBG has offered a number of relative advantages compared with the local development programs of the 1960s.

First is the simplification of the process by which funding could be obtained; only a brief formal grant application was now required, rather than the lengthy documents that typically had been associated with categorical grants. Although the Nixon administration would have preferred no review at all, HUD was allowed a short period of approximately 2½ months (75 days) in which to review and veto an application if the proposed local program failed to meet federal objectives (Dommel et al., 1982, p. 29). As opposed to the earlier categorical programs, the simpler application and the brief review period tended to provide greater assurance that local jurisdictions would receive funding unless their applications were seriously out of step with program guidelines. This made for more certainty and predictability regarding federal resources that would be available annually for planning.

Second and probably the most significant advantage to local officials was that they could now draw on a single block grant instead of multiple categorical grants in deciding the program priorities they would pursue. Rather than having to adapt their own spending plans to funds that were channeled through separate categorical grants each restricted to a specific use, local planners now could decide their own priorities within the limits of the consolidated funds available to them. This made for greater flexibility in local planning, as well as a greater degree of independence from federally established objectives. Another source of flexibility was provided by eliminating requirements for local matching contributions as a condition for grant receipt. In the past, there had been complaints that categorical grants that were premised on smaller local matching funds were a great temptation to tailor local planning to capturing "cheap" federal dollars (Frieden & Kaplan, 1975, p. 240), rather than to pursuing actual local priorities. Consequently, the shift from categorical to block grants and the removal of matching requirements were justified as incentives to localities to plan more independently and more creatively—in short, as incentives to local planning innovation.

These characteristics made CDBG seem a promising program to many local officials. However, in important respects CDBG proved to be quite problematic for public officials in older big cities and for the urban poor in general. For these groups, problems arose particularly with regard to the targeting of funds both between jurisdictions and within jurisdictions.

Targeting Between Jurisdictions

The adoption of CDBG involved new eligibility criteria for receipt of urban aid funds, which introduced a considerable spreading of such funds. Grants under CDBG would go by formula to all cities over 50,000 in metropolitan areas, to urban counties over 200,000, and on a discretionary basis to small cities inside and outside of metropolitan areas. Besides the numerous small cities that now likely would apply for discretionary grants, this meant that even without applying, various metropolitan jurisdictions that usually had not competed for categorical grants now were entitled automatically to funds from the CDBG; among them, many were prosperous suburban cities and counties. In addition, during the early 1970s new metropolitan area definitions by the Nixon administration's Office of Management and Budget (OMB) added almost 45 new metropolitan areas, many in the South. Altogether the combination of these several factors resulted in two funding shifts: a regional shift from jurisdictions in the Northeast and Midwest to the Southern Rim and, within any region, a shift from aging central cities to suburban and nonmetropolitan jurisdictions (Frieden & Kaplan, 1975, 1976).

Within this expanded range of eligible jurisdictions, the CDBG formula for distributing funds further tended to spread CDBG money away from previous major recipients of urban aid, particularly aging industrial cities in the Northeast and Midwest (Dommel et al., 1982, p. 41). The formula was based on total population, extent of overcrowded housing, and poverty population (which was given double weight). Because Southern Rim communities are often characterized by high levels of poverty, the double weighting of poverty and its typically high correlation with overcrowded housing both worked to the advantage of such communities in the CDBG formula. Moreover, both poverty and overcrowding as measured in the formula correlated closely with total community population. Thus the rapid growth of many southern and western cities meant that the formula's composition would operate substantially to their advantage as against cities of the Northeast and Midwest, which were generally undergoing serious population decline (Deleon & LeGates, 1976).

Targeting Within Jurisdictions

Unlike earlier major urban grant programs such as urban renewal or Model Cities, which had focused on particular designated areas of the community, the CDBG can cover any extent of the recipient community, up to and including the whole of it. Given that option, numerous communities have tended to distribute their CDBG funds to "neighborhoods less seriously distressed than those that had been the primary beneficiaries of the earlier categorical grants" (Henig, 1985, p. 184). In practice, the program functions mainly as a package of physical improvement grants that are applicable to a broad spectrum of area types, no longer predominantly those inhabited by the poor. In relation to central cities, CDBG can be used in a variety of locations ranging from older business districts to historic preservation areas or declining middle-class neighborhoods. Clearly the objective of targeting urban program funds to areas of poverty is no longer a high priority.

Contributing significantly to the shift away from targeting to the poor was the fact that the program emphasized physical renewal; 100% of CDBG funds could be used for physical improvements if a community so desired, whereas no more than 20% could go for social services (Frieden & Kaplan, 1976, p. 29). This was a direct reversal of the human renewal spending priorities that had characterized the Model Cities program; as of 1971 almost 70% of spending in Model Cities areas was for social services, with emphasis on health, education, and manpower development (Levitan & Taggart, 1976, p. 172). The new concern was for providing "hardware" through construction of public facilities, in particular, through the completion of "almost 1000 urban renewal projects and 400 related neighborhood development programs" (Frieden & Kaplan, 1976, p. 30). This amounted to an updating and modification of the earlier urban renewal program from one mainly focused on large-scale clearance and renewal to one emphasizing more modestly scaled rehabilitation and related local public works, especially in residential preservation or gentrification areas. An option for the funding of local social services ("software") was still available, but only on a limited basis.

Ultimately these shifts in the focus and character of urban aid associated with CDBG can be traced to a basic ambiguity incorporated in its authorizing legislation. As identified there, the foremost objective of the program is the provision of decent housing and a suitable living environment to persons of low and moderate income; local jurisdictions are required to certify that projects benefiting such persons are given maximum feasible priority. However, the legislation also points out two other major program objectives: the prevention or elimination of slums or blight and meeting community development needs that are deemed particularly urgent (Rosenfeld, 1980). In effect, the act combines the emphasis of the Model Cities program (on targeting to the poor) with that of urban renewal (on eliminating slums or blight) and includes an "urgent needs" clause to cover other contingencies. Such ambiguity is similar to that of the original urban renewal program of 1949, which included provisions both for public housing for the low-income residential community and for commercial renewal for the business community. In that case as well, ongoing program development and implementation increasingly involved a shift away from the interests of the poor.

Local Planning Under CDBG

Among the original justifications offered by President Nixon in the early 1970s for shifting to a block grant approach in community development was that governmental performance in community planning could be improved by reversing "the flow of power and resources" to Washington and directing it back to the states and local communities (Conlan, 1988, p. 11). The system of categorical grants was criticized as being uncoordinated, wasteful, and overloaded with too many inflexible federal requirements, thereby causing administrative confusion and distorting the priorities of state and local governments. All of this was held to be unnecessary and avoidable because these governments were said to be capable of effectively deciding their own needs and managing programs that would meet their own priorities (U.S. ACIR, 1977). CDBG's premise was that placing decision-making power in the hands of the localities would lead to more effective, less wasteful planning of urban development. But to what extent did this actually happen? What in fact were the nature and results of the CDBG local decision-making process?

As actually implemented, one distinctive feature of CDBG was that it tended to encourage the politicization of the local decision process. Under the programs of the War on Poverty, city councils generally had left the planning process in the hands of the city's "categorical agencies" (such as the Community Action Agency or the City Demonstration Agency) due to these agencies' experience in local planning and administration of formally categorical programs like CAP or Model Cities. However, in the case of this block grant program there was now a different attitude. Compared with earlier programs that had been targeted explicitly to poverty neighborhoods, the CDBG program was viewed as bringing into the community new money that was in a sense "up for grabs" to a broadened arena of competing organized local interests; this contributed to a much more politicized process as reflected in the crucial activity of developing the city's grant application to HUD. Thus, in the case of CDBG, what essentially characterized the planning process was not the exercise of neutral technical rationality but an unmistakably political type of activity: "*bargaining*—among elected officials, citizens, city bureaus, and neighborhood agencies—that was the most prominent feature of local decision making" (Kettl, 1979, p. 445; emphasis added). Elected officials involved in the process typically included the members of the city council as well as the mayor.

Effectiveness of Planning Under CDBG

Overall, this broadly inclusive bargaining process produced "scattered, short-term, neighborhood-based projects." Within any given community, the shift to CDBG "significantly increased the number of areas receiving funds" (Kettl, 1979, p. 446). Decision making was essentially distributive, especially to politically influential neighborhoods, producing a disjointed scattering of projects around the city. CDBG projects tended to be small scale, such as spot rehabilitation of housing within a particular neighborhood area or limited public works projects (such as construction of small park or recreation areas), as well as street improvements. The limited project scale was due in part to recurring uncertainties over congressional funding, as well as to frequent changes in HUD eligibility requirements. These considerations, together with concern for reelection every few years, pressed community officials to produce quick, visible local program results.

In response to these problematic characteristics of the local CDBG, a number of sharp criticisms have been leveled at the program. One is the relative lack of integrative planning, as a result of which CDBG has functioned as a collection of unrelated public works that were based less on deliberate planning than on politically driven responses to a mix of diverse local interest group demands. Cities made little attempt to prioritize needs or to establish relationships between different projects. Instead local decision makers "relied on developing projects with immediate appeal for identifiable constituencies. The local plan consequently was no more than a collection of individual projects that took little account of national program objectives" (Kettl, 1979, p. 451).

In relation to these objections, a frequent criticism has concerned the CDBG program's failure to target clearly the needs of the poor within communities. Although the authorizing legislation emphasized the importance of giving maximum feasible priority to the poor, it also referred to options for pursuing goals such as elimination of blight or other local development needs considered urgent. Yet if improvement of the poorest areas of the community had first priority, did this not imply some limits on the pursuit of other community development goals? Despite this, there was evidence that "local communities were in fact diverting funds away from the poor," in some cases constructing facilities such as tennis courts or marinas and in one case upgrading "a road used by people going to a country club" (Kettl, 1980, p. 25). This dispersion of CDBG projects away from concentration on poor neighborhoods follows from the nature of the CDBG's highly politicized local planning process, in which planning decisions respond to the balance of influence between different interest groups from diverse locations—socially as well as spatially—in the community.

These aspects of local CDBG planning left relatively unresolved some significant issues of program agenda setting and accountability with regard to the use of CDBG resources. Reflecting on this, Kettl (1979) asked whether cities could "be trusted with great discretion over large sums of money?"—a basic feature of the block grant approach to funding community development. Or was there a need for tighter federal scrutiny of local program decisions in the CDBG program? His conclusion was that "the cities appear politically equipped to deal with scattered, short-term, neighborhood-based problems," but that problems that demand more sustained commitment seem to be "beyond local political capacity" (p. 451).

Those problems include broad-scale community needs such as long-term economic development and provision of affordable housing, as well as the general need for improved targeting of program resources to the poor. The underlying difficulty is that local resource allocation under CDBG is too much dictated by pressures for quick results aimed at satisfying a wide variety of organized constituencies: "The time horizon of local officials is too short and the demand for widespread distribution of benefits is too great to allow local governments to launch a concerted attack on these problems. Attacking national problems through federal grants demands greater federal control over local decisions" (Kettl, 1979, p. 451).

Thus although CDBG may be adequate for dealing with scattered problems of relatively small scope and short duration, its highly decentralized program structure is inappropriate to the solution of community problems of broad scope or problems demanding long-term financial and political commitment.[1] Contrary to the initial predictions of its advocates, CDBG has shown serious limitations in responding effectively to some of the deeper and more fundamental economic and social issues in contemporary community development.

Politics, Planning, and "Triage" at the Local Level

Although CDBG tended to encourage piecemeal action rather than integrated planning, that does not mean that there was no pattern to its effects as a politicized planning system. I noted earlier CDBG's deemphasis of targeting program resources to poor neighborhoods. Underlying this deemphasis is a combination of political considerations and planning strategies that often has affected the thrust of program choices made at the local level. Usually, local officials have tended to be more responsive to middle-income neighborhoods because they are "generally more organized and politically active than lower-income areas." In a study of 137 cities it was found that "[b]etween 1975 and 1979, the number of low and moderate income census tracts receiving CDBG funds increased by only 30%, as against a 50% increase in the number of wealthier tracts receiving benefits" (Henig, 1985, pp. 184, 185).

It is not surprising to find local political elites paying this type of attention to their better organized, more influential constituencies. Moreover, HUD— as the federal agency that is authorized to oversee CDBG program implementation—has been relatively permissive in allowing broad discretion to local officials administering the program. But political organization and influence are not the only significant considerations taken into account by local officials. In their view, directing funds into transitional neighborhoods (especially aging middle-class neighborhoods in process of physical decline) rather than those in the worst condition promises in the near term to produce more positive results. In particular, it appears much more likely both to produce visible local improvements and to benefit the city's budget, thereby appreciably boosting these officials' public image.

For many local officials administering the CDBG program, this type of calculation is made informally, as a type of common sense approach to program implementation, rather than on the basis of an explicit plan. In effect it takes the 1960s planning theme of giving priority to the neediest areas ("do the worst first") and changes it to giving priority to neither the worst nor the best, but to the "graying," run-down but not yet slum areas of the city in between while they can still be saved. The resulting program is an urban version of a system of administering medical aid to battle casualties that originated during World War I, known as *triage*, or treatment based on a standardized three-part classification. In general, the best cases were those only lightly wounded, able to survive with little or no treatment (healthy neighborhoods); the worst cases were those too far gone to benefit from treatment (analogously, this judgment is applied to fully deteriorated neighborhoods); and the intermediate cases were those whose wounds were not regarded as fatal but sufficiently serious to receive substantial treatment (partially deteriorated, or transitional neighborhoods). Based on the triage model, transitional areas would be given priority for public projects aimed at helping reverse their decline.

Included in the debate on the legislation that created the Community Development Block Grant program were several proposals of significance for the later implementation of CDBG. One of the more explicitly stated provides a clear elaboration of the triage model in its application to older declining cities. The typical older city was to be divided by program planners into three basic area types for differential treatment. First were the conservation areas, essentially healthy and able to attract private investment, hence requiring only minimal city attention such as rigorous code enforcement to maintain property standards. At the opposite pole were the depletion areas, seriously deteriorated and already suffering a combination of private disinvestment (including redlining by local lending institutions) and public withdrawal of services.[2] In these areas, the city now would discourage any reinvestment; furthermore, abandoned land and properties would be left unused or "banked," until the area was deemed "ripe" for "complete redevelopment." Finally, in a parallel to the transitional type in the triage model were the redevelopment areas, somewhat run-down and showing potential for decay but not yet badly deteriorated, hence deemed worthy of "major public investment" for the sake of "longterm revitalization" (Pratter, 1977, pp. 81-87).

In the redevelopment areas local government would agree to make strategically placed public infrastructure investments (e.g., in streets and sidewalks, plantings, lighting, parking, water, and sewage) that would indicate clearly to investors its strong commitment to a project, as well as contributing to the project's technical implementation. The intended result would be to encourage private investment of a magnitude at least several times larger than that undertaken by the city (whether large-scale commercial investment by development firms, or quite usually in CDBG, accumulated small investments in housing rehabilitation). In effect this combined the triage model with the concept of public sector leveraging of substantial private investment (Lachman, 1975). By the late 1970s, such an approach gave evidence of a decided shift from the War on Poverty concept of investing public resources in the provision of social services for human renewal to an emphasis on investing public resources to stimulate private investment in physical renewal, thus changing not only the major objectives of urban policy but also the basic strategy for their achievement.

CDBG: The Changing Role of Citizen Participation

With the advent of CDBG there occurred not only a major shift in the targeting of urban program resources but also significant change regarding the role of citizen participation and the procedures associated with it. Much was made by the Nixon administration of the notion that grant decentralization through the programs of a New Federalism would mean a return of power to the local level. As the president framed the issue in a radio address:

> Do we want to turn more power over to bureaucrats in Washington in the hope that they will do what is best for all the people? Or do we want to return more power to the people and to their state and local governments so that people can decide what is best for themselves? (Clark, Iglehart, & Lilley, 1972b, p. 1911)

This argument proved persuasive in that a certain amount of support for passage of CDBG legislation "was based on the belief that decentralization would stimulate greater citizen participation and a more democratic process through which development policies would be shaped" (Henig, 1985, p. 185). Yet, contrary to these expectations, when President Nixon took office in 1969 his administration quickly acted to minimize the role of citizen participation, reducing its function in the Model Cities program to a purely advisory one, "thus protecting the authority of the local elected officials" (Dommel et al., 1982, p. 32).

Consistent with this, in its 1973 proposal for the CDBG—a major block grant program that stands as a prime example of the New Federalism—the Nixon administration required nothing more than publication of a community development plan by local officials 60 days before it became final, thereby allowing a brief period for analysis and comment by interested parties. There were no provisions for public hearings and no guidelines as to what official action would be required in response to citizens' comments. It was only due to congressional recommendations that the final version of the CDBG legislation mandated at least two public hearings before a grant application could be submitted to HUD by local officials. Although the legislation also required that citizens should be given an opportunity to take part in developing the application, it did not spell out by what procedures, and there was no mention of citizen participation in the execution of the program (Dommel et al., 1982).

The contrast with earlier urban programs, particularly those of the War on Poverty, is striking. In those programs federal officials devised explicit procedures for citizen participation; in the case of CDBG, HUD guidelines provided no suggestions as to such procedures. This was not an accidental oversight. When critics pressed HUD to specify the basic structures and procedures required to ensure local citizen participation, the agency refused to do so (Morris, 1976). It was now apparent that under the New Federalism, HUD's conception of participation had become a drastically restricted one. Essentially, after program planning was completed, local government simply would inform citizens about the location, purpose, and amount of planned CDBG expenditures involved. As a concession to some remaining vestige of citizen participation, residents would be allowed only then to express their views on the projected plans at a few public hearings.

In part these limits on resident involvement can be traced to the hesitancy about vigorous participation that is evident in the authorizing legislation. The Housing and Community Development Act of 1974 stipulated that citizens should be provided adequate opportunity to participate in shaping the locality's application for a community development grant. However, it also stated that the requirement for citizen participation could not be interpreted as restricting local government's authority over the development of the application (Morris, 1976). Thus under the rhetoric of adequate opportunity for participation, the option was left open to local government to do little more than conduct several hearings pro forma with no way specified to ensure that citizens' comments and suggestions actually would be taken seriously and incorporated in the local submission to HUD (Kettl, 1979).

Such weakening of the citizen participation process had its most serious consequences for low-income and minority residents of grant-receiving cities. The fact that the CDBG allowed for wide expansion of the boundaries of project activity tended to contribute further to these developments. In the Community Action and Model Cities programs, access to the planning process had been targeted explicitly to the residents of designated poverty neighborhoods. But under CDBG, the program could operate in any part of the city, so that citizen participation likely would involve a greatly broadened spectrum of interests and an expanded range of demands, many of them far removed from the focus of the neediest areas.

CDBG: The Changing Structure of Participation

In CDBG's expanded context for participation of competing urban interests, one advantage that residents of lower income areas initially seemed to have was that many of their communities already had been organized for participation during the War on Poverty. As long as those participation structures were not abolished outright by local chief executives, they were still available to be carried over or replaced with new structures. However, a 6-year case study begun in 1975 indicated that even in cities in which earlier structures of participation (such as neighborhood development committees or Model Cities councils) were retained, under CDBG those structures now tended to receive decreased attention due to the emergence of numerous competing constituencies (Dommel et al., 1982).

Changes in citizen participation were most pronounced in cities that chose not to retain earlier participation structures, instead replacing them with the minimal arrangements for participation associated with CDBG (typically public hearings or advisory councils). In those cities there was a distinct shift from representation of a limited number of target neighborhoods to representation of a wide array of city interests. This expanded arena for decision making was entered by many other local area interests across the spectrums of race and class. In addition, it now could include a variety of predominantly middle- to upper-class citywide organizations, from civic groups such as the League of Women Voters to business associations such as the Chamber of Commerce. Under these circumstances, the perspectives of representatives of low-income minority areas have tended to receive a relatively low level of priority in the CDBG decision process.

Furthermore, CDBG channeled federal program resources directly to the local chief executive, thereby fostering a general elevation in the power of the central urban executive (mayor or city manager), which included increased influence over the conduct of citizen participation. Thus in the CDBG program, the highest level of participation was now contingent on executive appointment to a citywide council limited to a basically advisory role; in the antipoverty programs it had been by local area election to neighborhood councils that could strongly influence program operation. This shift was accompanied by the increased dominance of executive staff in program decision making with citizen advisory bodies acting largely to legitimize staff decisions.

All of this enhanced the importance of executive decision makers and their technically versed planning staffs or technocrats, much as the early 20th-century reform movement had legitimized the role of strong mayors and professional bureaucrats as policy makers and administrators in local government. In general, with regard to the relation between citizen participation and social status, the new patterns tended to facilitate participation in the planning of local community development by upper socioeconomic groups and community decision elites and the marginalization of lower status groups (Dommel et al., 1982).

In each of these changes can be seen a certain repetition of the patterns of early 20th-century urban reforms that worked to reduce the political participation of lower income groups. Like those earlier reforms, the changes in participation brought by CDBG tended to reduce the participatory access gained by low-income community groups that had managed to organize with the support of programs of the War on Poverty. Through federal-urban channels, those antipoverty programs had nurtured the emergence of a type of neighborhood organization politics in poor minority neighborhoods, to which traditional White-ethnic-dominated machines long had declined to give more than token support. This new neighborhood-based politics was viewed by political professionals as amounting to the creation of a new machine politics tied to and encouraged by federally funded local program agencies such as the Community Action or local Model Cities agencies.

Many politicians still associated with the remnants of the traditional urban machines saw the new antipoverty agencies as a direct challenge to what remained of their own political organizations. Indeed, the Model Cities program explicitly had encouraged a parity model of officially equal status in bargaining between representatives of targeted poverty neighborhoods and city hall. The CDBG program's deemphasis of that model tended to be particularly attractive to old-line party politicians who feared that such influence for inner-city neighborhood representatives would undermine further their own already weakened position in local politics. Thus, with the establishment of CDBG, changes in community participation structures occurred that tended to complement changes in program funding in the shift from categorical to block grant. In short, the new CDBG citizen participation practices amounted to an organizational disempowerment of urban poverty areas.

Ultimately, these changes in the structure of participation worked together with the fiscal defunding of those areas caused by the "spreading effects" of CDBG distributive patterns to produce a sharp deconcentration of grant resources away from the poorest, most disadvantaged urban neighborhoods (Frieden & Kaplan, 1976, p. 43). This was but one in a series of such steps initiated by the Nixon administration as part of its developing design for a New Federalism, left in place by the caretaker Ford administration (installed after the Watergate scandal resulted in Nixon's resignation and early departure from the presidency).

CDBG had a number of weaknesses with regard to providing benefits to low- and moderate-income urban residents. Yet despite this it remained the only major urban program that was in any way directed to these groups and their neighborhoods. In the next sections I trace the CDBG program under the Carter administration that followed the Nixon-Ford period, examining shifts in implementation of CDBG in the context of that administration's attempts to produce a clearly defined urban policy.

THE CARTER ADMINISTRATION
AND THE NEW FEDERALISM

Viewed in retrospect, the Carter administration was essentially a transitional one, characterized by change and ambiguity in its urban policy. This administration marked the end of an era of expansionary urban policy and the beginning of significant reductions in urban program spending. It also involved the adoption of policy initiatives emphasizing fiscal incentives such as the use of federal grants and loans to stimulate urban business investment in cities rather than the expansion of spending for redistributive social programs. Its shifting, transitional character was in part because the problems of the economy were assuming a new and unaccustomed form (stagflation) that placed sharp constraints on the urban policy initiatives with which President Carter initially identified himself. At the same time, the Carter administration's policy instability could be traced also to its relatively weak political authority, resulting in a general failure to establish firmly its policy agenda with the Congress.

By the 1970s changes in the American economy that had been at work for some time began to make themselves sharply felt in the occupational and community lives of broad segments of the population. These economic developments continued to exert their effects during the rest of the decade, ultimately raising serious policy issues for the Carter administration, particularly in large older industrial cities in the Northeast and Midwest where important segments of the Democratic constituency still were located. One set of changes—the transition from an economy centered around industrial manufacturing to a broadly service-based postindustrial economy—was reflected in the changing occupational structure of the urban labor force. Increasingly, the labor force of large central cities was less likely to include sizable percentages of blue-collar workers with relatively well-paying, secure factory jobs. Now growing numbers of former blue-collar workers were experiencing protracted unemployment or finding work in a variety of low-wage service occupations (Kasarda, 1985).

Another set of changes with significance for urban policy were related to the declining rate of economic growth. Vigorously challenged by foreign competition in major industries such as autos and steel and buffeted by the shocks of energy shortages, the American economy became characterized by slow growth. Stagnation in the national economy was reflected in rising unemployment and complicated by the growth of inflation—hence the composite term *stagflation,* which ultimately became a political label for economic problems that came to a head under the Carter administration (Dolbeare & Edelman, 1985), with particular intensity in older industrial cities.

Three overlapping policy stages relevant to urban development policy can be identified as emerging over the course of the one-term Carter administration. First was a stage of policy initiation for economic and urban recovery, beginning in 1977 and continuing during the following year, which included antirecessionary stimulus to the general economy and strong emphasis on the targeting of federal aid to distressed cities. Second was a policy review stage leading to the explicit formulation of a Carter urban policy, reflected in the March 1978 publication of a proposal for a *New Partnership* involving relationships of close cooperation between federal, state, and local governments and between business, government, and community groups in carrying out urban policy. Last was a contractionary policy stage beginning in 1979, involving reductions in federal programs that had expanded during the preceding stages and a shift to a narrowed structure of New Partnership.

To help revive a sluggish economy and reduce unemployment, the Carter policy initiatives included a major economic stimulus package of fiscal assistance, public works, and job training, as well as initiatives for improved targeting of the CDBG program to economically impacted areas. In addition, a new program of aid specially earmarked to help stimulate private investment in distressed cities was established (known as the *Urban Development Action Grant* or UDAG program). During this stage the Carter administration tried to reintroduce certain aspects of the War on Poverty such as distress-sensitive aid, within the limits allowed by the restructured system of intergovernmental relations now known as New Federalism. However, although HUD under Carter emphasized the need for targeting of federal aid to distressed communities, it ultimately left the specific determination of program priorities to local decision makers in both the CDBG and UDAG programs owing to the strongly decentralized structure of decision making established by the New Federalism.

Through UDAG the Carter administration increasingly came to support the creation of public-private partnerships for urban revitalization, providing federal subsidies to joint development ventures between local governments and private for-profit investors. UDAG was an economic development program targeted explicitly to economically distressed cities, serving a function for which neither general revenue sharing nor CDBG were designed (Jacobs & Roistacher, 1980). In contrast, as a community development program, CDBG has been much more diffuse than was UDAG, both in its objectives (including for example, housing rehabilitation, neighborhood facilities, and open space, as well as economic development) and in its territorial scope (embracing not only large metropolitan area cities but also large suburban counties and small nonmetropolitan cities). In effect then, diffuse CDBG and targeted UDAG constituted roughly complementary core elements of urban-relevant policy during the Carter administration.

Overall, the first stage of the Carter administration's urban policy combined short-term antirecessionary fiscal and public employment policies with long-term community development (CDBG) and urban economic development (UDAG) policies, plus the public sector employment and training program known as CETA. This was the most substantive stage in terms of its developmental initiatives, as well as most interesting in regard to basic issues such as program targeting and citizen participation. The major focus in my discussion of this stage will be on developments in the CDBG program, which has continued in existence since the mid-1970s as the centerpiece of urban-relevant policy under the New Federalism.

Economic Stimulus and Targeted Aid

On coming to office in January 1977, the Carter administration faced the aftereffects of the economic recession of 1974-1975. These included an official unemployment rate of 8.5% in 1975 and 7.7% in 1976 and the beginnings of a resurgence of the poverty rate, which rose from 11.2% in 1974 to 12.3% in 1975 and still remained above 11.5% in 1977 (U.S. Bureau of the Census, 1981, pp. 380, 446). Thus issues of unemployment and poverty and their concentration in large urban communities were placed unavoidably on the national agenda once again. Although this was not the most favorable time for taking on new commitments, President Carter nevertheless pledged his strong support for dealing with urban problems.

Among his early policy initiatives he submitted legislation to Congress aimed at providing fiscal aid to localities, stimulating economic recovery, and reducing local unemployment (Palmer, 1984), as well as seeking closer targeting of federal employment aid to jobless individuals with greatest need. Basically, this legislative package extended several existing grant programs, including a 1-year $2-billion extension of a counterrecessionary program of fiscal assistance to distressed local governments to help them "avert layoffs, maintain services, and avoid tax increases" (Judd, 1988, p. 353) and a 2-year $4-billion extension of a public works program that in part targeted funding for construction jobs to distressed localities in states experiencing high rates of unemployment—at that time defined as more than 6.5% (Palmer, 1984,

p. 37). Both were familiar countercyclical programs aimed at combating the relatively short-term effects of a downturn in the business cycle. Also in the package was a program of potentially greater significance in dealing with emerging long-term problems of structural unemployment based on the lack of required job skills for an economy increasingly involved with the provision of services, both public and private.

This was the 1-year reauthorization of the Comprehensive Employment and Training Act and its orientation more specifically to the hard-core unemployed (modified in this way, it was extended again in 1978 for another 4 years). This measure provided funding for over 650,000 public service jobs, thus substantially expanding the payrolls of many fiscally hard-pressed local governments. With this reauthorization, the focus now was on providing entry-level public sector jobs and training for persons with little prior work experience, rather than allowing the diversion of CETA funds to ensure employment for the regular public workforce—a practice that previously had earned the program much criticism (Palmer, 1984, p. 37). Thus under the Carter administration CETA came to be redefined as a redistributive public sector employment program, as opposed to its essentially distributive implementation under Nixon and Ford. However, the definition of CETA as a redistributive program apparently gave it a character not readily acceptable to the conservative administration that succeeded Carter; in 1981, the first year of the Reagan administration, the program was eliminated.

COMMUNITY DEVELOPMENT UNDER CARTER

In contrast to CETA, the CDBG program (the other major block grant created during the Nixon-Ford period) was retained and still remains significant as the major federal program of relevance to urban area development since beginning operation in 1975. In its first years, however, such continuity by no means was assured, owing to the pronounced tilt in its distributional formula in favor of growing versus declining communities. Under the 1974 Housing and Community Development Act, which established the CDBG, distribution of its block grant funds was based on the size of community population, the amount of overcrowded housing, and the extent of poverty (double weighted). Based on that formula, cities that were losing population (as in the Frost Belt) increasingly would be disadvantaged, but cities experiencing rapid population growth (such as those in the Sun Belt) would qualify for more money every year. In addition, the poverty factor in the formula also benefited southern cities because most of them had low wage scales and high poverty rates for their minority populations (Palmer, 1984).

In view of these features of the CDBG formula, by the end of 1975 there was concern that major Frost Belt cities would receive a progressively smaller share of community development funds over time. Divergences in demographic trends between older northern cities and younger, fast-growing southern and western cities meant that industrial Frost Belt cities were destined to undergo continuing declines in CDBG assistance levels. Over the short term, such cities were protected for 3 years by a "hold harmless" provision in the original legislation, which assured that they would receive no less than the annual average previously received under the categorical programs that had been consolidated into the block grant. However, as of 1978, this provision no longer applied, so that by the late 1970s many northern industrial cities would be facing considerable reductions in federal program assistance (Dommel et al., 1982).

For the Carter administration, with major constituencies in distressed industrial cities, it was urgent that aid be targeted to needy localities rather than widely spread about as in general revenue sharing or in the CDBG program as created under the Nixon administration. (Although in 1968 almost two thirds [62%] of all federal urban grants went to cities over 500,000 population, by 1975 less than half [44%] went to cities of that size. See Palmer, 1984, p. 36.) To correct for this, the proposed economic recovery program was designed to refocus funds on communities with high rates of unemployment and poverty. In particular, legislative changes contemplated by the administration would have revised the CDBG formula to channel substantially increased federal aid to older big cities in the North.

Pressing hard for these changes, in 1977 the Carter administration persuaded Congress to reauthorize the CDBG program and asked it to amend the distributional formula to target federal aid to distressed areas in greatest need. Pursuant to the findings of a Brookings Institution study, CDBG funding would be based on a formula incorporating several factors: a city's growth lag relative to average urban population growth (weighted 20%), thus favoring cities losing rather than gaining population; poverty rate (weighted 30%); and age of housing stock (weighted 50%), another factor favoring older big cities (Palmer, 1984, p. 38).

The upshot was that although this would protect the allocation shares of older central cities, it significantly would reduce the share going to newer urban areas (such as growing urban jurisdictions outside metropolitan areas) from what they would have obtained by the late 1970s under the old formula.

To prevent this a compromise was struck in the mode of classic "formula politics." Instead of simply replacing the old formula that would have hurt large northern cities by the new one which favored them, "a dual formula system" was established "in which each city would get the larger of the two amounts computed with both formulas" (Hays, 1985, p. 215). This roughly cut in half (from 27% to 14%) the possible losses of central cities (the major group in the CDBG system) while largely preserving funding gains of non-metro area jurisdictions and slightly improving funding gains of metro satellite cities and urban counties (Hays, 1985, p. 213).

CDBG as Hybrid: Program Ambiguity

Despite its success in shifting community development block grant monies to Frost Belt cities, the Carter administration's goal of specifically targeting aid to areas of greatest need was not achieved so readily at the level of any given city. Although the administration succeeded in directing CDBG funds to communities with high levels of economic distress, it was less successful with regard to targeting within the recipient communities. Once again, as in the case of Model Cities Planned Variations, issues of local implementation stood in the way of achieving national policy priorities. In part this could be traced to the mixed composition of the original legislation.

As enacted, the CDBG program approved in 1974 could be described as a hybrid combining several different elements. On one hand, in order to gain the support of the urban liberal bloc in Congress, it advanced the objective of *social targeting* of program benefits to low- and moderate-income persons— to be pursued as a maximum feasible priority, hence presumably as the program's leading goal. It promised that "decent housing and a suitable living environment" plus "expanding economic opportunities" would go "principally" to lower income persons (Rosenfeld, 1980, p. 212; U.S. Congress, 1974).

On the other hand, reflecting the objectives of the Nixon administration's New Federalism, CDBG also incorporated the goal of decentralized decision making as the basis of program implementation. This meant allowing considerable discretion to local officials regarding just how much emphasis to place on social targeting, particularly as against the program's second basic goal, stemming the spread of slums and blighted areas largely through neighborhood revitalization, involving a mix of local commercial renewal and housing rehabilitation (often in the form of gentrification). A third program goal— meeting other community development needs deemed urgent—although less frequently invoked in program implementation, on occasion has been used to justify construction of public tennis courts or marinas. Thus the different objectives associated with social targeting and these other goals were placed in a continuing competition (Dommel et al., 1982; Rosenfeld, 1980).

The result was a serious ambiguity in program emphasis, which set the context for tensions later encountered in the program's implementation. The 1974 Housing and Community Development Act did not take effect until 1975. By then, congressional investigation of the Watergate episode had led to Richard Nixon's resignation from the presidency. Hence implementation of the community development act was not begun until the Ford administration, which succeeded Nixon. Yet in its two years in office, that administration "did not adopt any clear policy on how to implement the social targeting language" that had been included in the 1974 act (Dommel et al., 1982, p. 243). In particular, the absence of a clear definition of maximum feasible priority meant that CDBG program priorities would remain uncertain.

Targeting CDBG

Ambiguity regarding program priorities seemed to end with the inauguration of the Carter administration in January 1977. Reports of a decline in program benefits to low- and moderate-income citizens during the preceding year led the newly appointed HUD secretary to commit the agency to placing strong emphasis on the social targeting objectives of the CDBG program (Dommel et al., 1982). Shortly thereafter, new HUD regulations restated the priority of social targeting to lower income persons, allowing no more than 25% of a community's grant for projects that satisfied either neighborhood-revitalization or urgent-need goals. This made it clear that at least 75% of program benefits would go principally to lower income persons. Using this "75-25" rule, the program's major goals could be pursued with much less ambiguity, to the relief of civil rights groups and many neighborhood organizations. However, the new regulations soon triggered strong opposition from local government officials and influential members of Congress.

In response, HUD agreed to a compromise that left matters still somewhat cloudy. In evaluating a community's application for a CDBG grant, HUD henceforth would take into account more explicitly the level of benefits planned for lower income households. On the other hand, however, as a major concession to the discretion of local officials, there would be no clear-cut federal statement of leading program priorities. This again left the issue to local public decision makers acting within the particular structure of political influence (community power structure) that prevailed in their respective localities. In many urban communities, such an arrangement tended to work against the primacy of social targeting to lower income groups. Thus even with a moderately liberal administration in office, execution of policy priorities on behalf of the poor was impeded by the strongly decentralized structure of implementation characteristic of the New Federalism.

Another type of targeting policy developed by HUD under the Carter administration was *geographic targeting,* directed mainly at improving implementation of neighborhood renewal. This was already an important objective of local public officials because it fostered visible distributive results that paid off at election time. Here the Carter administration expressly made neighborhood revitalization a key program goal and articulated several related planning objectives that had been lacking in the relatively laissez-faire approach to local CDBG programming under the Nixon-Ford administration. Instead of scattering unrelated projects throughout the community, a more concentrated and rational use of block grant resources would be encouraged henceforth. The underlying objective was to support strategic planning that would efficiently "stabilize and upgrade residential areas affected by blight and deterioration" (*Federal Register,* 1977, p. 56465).

However, in thus emphasizing the importance of efficient neighborhood revitalization to the success of the CDBG program, the Carter administration over time tended to substitute geographic targeting, particularly to aging middle-class areas, for social targeting to poverty neighborhoods. Initially this was a concession to the locally based political obstacles that blocked the administration from continuing the commitment to social targeting earlier pursued by the War on Poverty. But eventually it also could be seen as consistent with the administration's own growing emphasis on using public expenditures to leverage private investment in those areas considered economically viable for investment. This approach was associated with a heightened interest in fiscal efficiency as opposed to social equity, with fiscal efficiency becoming a major policy objective during the later years of the Carter administration (Smith & Judd, 1984).

Local Implementation and Impact of CDBG

A look back at major postwar urban programs shows that in contrast to CDBG, Model Cities and urban renewal both were targeted significantly to low-income areas, although with distinctly different effects. With its strong emphasis on social services, Model Cities consciously directed its benefits to poor persons located in inner-city neighborhoods. On the other hand, urban renewal often tended to displace such persons while most program benefits went to more affluent interests—frequently, business or professional occupants of new downtown office buildings or apartment towers in adjacent residential renewal areas. In comparison, CDBG has tended to locate most of its projects in the aging residential gray areas of central cities, which generally stand physically as well as qualitatively between seriously deteriorated lower income inner-city neighborhoods and the more affluent neighborhoods of the city's outer areas.

Focusing on the gray areas has meant that most CDBG benefits would flow to older middle-income neighborhoods seeking to preserve themselves from serious decline, largely through physical upgrading. In those cases where CDBG has undertaken projects in low-income neighborhoods, the results have been relatively diffuse from a social-targeting perspective. In addition, public works intended to improve the local physical environment (e.g., street repair, lighting, park improvements) often have provided "locational benefits" to commuters and other outside users of the area's streets and facilities as well as to the area's residents in general, rather than specifically to its neediest households (Dommel et al., 1982, p. 251).

Until the early 1980s, CDBG's emphasis on gray area upgrading through physical improvements contributed to the spreading of program benefits away from residents of lower income areas. By comparison, a combination of social services and housing rehabilitation better targeted to low-income neighborhoods could have provided greater benefit to those residents. However, CDBG regulations explicitly limited social service spending to no more than 20% of the grant allocation. In addition, determining which residential areas would receive federal aid for housing rehabilitation has depended on program decisions made at the local level. There, the thrust of program policy tended to work against lower income households because "many jurisdictions adopted grant and loan standards well above the low- and moderate-income level to keep higher-income residents in the neighborhood" (Dommel et al., 1982, p. 251). As a result, CDBG-funded housing rehabilitation and local public works frequently have had middle-class residential stabilization as its first priority; in areas of historic interest or favorable environmental features (parks, waterside locations), CDBG also has supported gentrification, converting older but still viable moderate-income residential areas to upper middle-class standards.

Overall then, as of the early 1990s CDBG had fallen short of fulfilling early HUD declarations of strong commitment to targeting benefits to poor inner-city residents. Its initial promises to provide decent housing and expanding opportunities for lower income persons were not pursued as top priorities. Instead, to the extent that it has been implemented as a neighborhood revitalization program, CDBG has proved of greatest benefit in stabilizing and improving older middle-class neighborhoods. From the standpoint of those areas and of local policy makers who see the retention of an urban middle class as essential to central city prosperity and stability, the program can be viewed as something of a success. In this it illustrates how a highly decentralized program structure can contribute to the blocking or diversion of redistributive national policy objectives by leaving their implementation largely to the interplay of established political interests at the municipal level.

As several insightful analysts have observed, politicized programs such as CDBG have tended to be most effective in advancing developmental objectives, although professionally managed public programs have better served redistributive policy objectives (Peterson, Rabe, & Wong, 1986). Owing to their decentralized nature, block grants tend to place program benefits in a relatively open local public arena in which they are subject to intense political competition. In comparison, categorical programs define a relatively narrow set of constituents at the local level and target benefits explicitly to them alone. Thus they tend to be more protective of the interests of disadvantaged groups as program clients. (By the same token, however, unless those groups become organized for effective participation as constituents of those bureaucratically managed programs, administrative paternalism can become the norm, with its own dysfunctional consequences.)

From the perspective of the vertical intergovernmental dimension, categorical grants encourage a structured set of interorganizational linkages between federal and local agency professionals dedicated to the nonpolitical administration of programs with redistributive objectives—thereby insulating program resources at the local level from acquisition by powerful competing interests. Such in fact were the objectives of federal Great Society planners during the 1960s, but the expanded categorical system fostered by their program initiatives became administratively weak due to excessive complexity and politically vulnerable due to overly close identification with antipoverty policies. By contrast, CDBG was a product of the New Federalism that was offered by conservative national policy makers in the 1970s as an alternative to the expanded categorical system. In particular, the decentralized CDBG program structure made available to urban middle-class interests a politicized local program arena in which they could achieve a relatively high level of access to resources previously targeted exclusively for the redevelopment of poverty areas. Ultimately this contributed to the diversion of the Carter administration's initial intentions to put CDBG to use in pursuit of a revived urban agenda on behalf of lower income groups.[3]

URBAN REVITALIZATION
AND THE NEW PARTNERSHIPS

The preceding analysis of the Carter administration's implementation of the Community Development Block Grant program has traced the effects of underlying ambiguities in the program, reflecting compromises that shaped the 1974 authorizing legislation. By the last half of Carter's term in office, under conditions of economic stagflation and a tax revolt spreading across the nation, these ambiguities increasingly tended to be resolved by abandoning earlier suggestions of an antipoverty aspect to CDBG. Nevertheless, despite these problems the Carter administration made special efforts to articulate an explicit policy as the basis for its urban program initiatives. This was a break with all previous practice of allowing much of urban policy to be the implicit outcome of other, basically nonurban policies (including highway, defense, agriculture, and federal mortgage guarantees, for example).

Carter's Partnership-Oriented National Urban Policy

Soon after assuming office, President Carter appointed a cabinet-level Urban and Regional Policy Group to assist him in producing a comprehensive urban policy. In March 1978 the group delivered to him a statement with numerous recommendations about existing urban policies and options for policy innovation. Based on these recommendations, he proposed a New Partnership of cooperating federal, state, and local governments working together mainly with the private business sector and secondly with local community groups to address long-standing urban problems. On a temporary basis he proposed increases in countercyclical fiscal aid to economically distressed cities. For the long term, the leading related policy initiatives proposed were:

a. Strong emphasis on economic revitalization of cities through encouraging private business to invest or to locate in older central cities. This approach called for strategically targeted provision of public incentives to stimulate private investment in distressed cities and the creation of local public-private partnerships to design and administer the development programs associated with these joint investments.

b. A national Urban Development Bank (Urbank) that would provide public investment funds to urban development projects in economically distressed central cities. The bank "would have the authority to provide more than $8 billion in loan guarantees to businesses and another $3.8 billion in interest rate subsidies" to lower investor's rates of borrowing (Harrigan, 1985, p. 387). Thus the bank represented the major organizational mechanism for delivering resources to implement the public-private revitalization strategy.

c. Urban impact analyses. These would evaluate the likely impact of all federal policies, whether urban or essentially nonurban, on "the fiscal conditions, economic development, and population and residential location patterns in cities" (Markusen, 1980, p. 107). Where adverse effects could be predicted, this presumably would bring the responsible federal agencies to make suitable modifications to avoid such impacts. However, critics questioned whether such knowledge alone would suffice to produce program changes where well-connected program constituencies such as the business establishment were prepared to "fight hard to preserve their publicly provided incentives" (p. 116).

With regard to Urbank, the institutional centerpiece of Carter's public-private development policy, intense interagency competition for its control led to a compromise that would have involved a rather complex interorganizational arrangement by which the new agency would be governed jointly by the Departments of Commerce, HUD, and Treasury. Because Congress did not respond favorably to this proposal, a subsequent administration proposal designated the Economic Development Administration (EDA) of the Commerce Department as the governing authority, but this was not successful either, owing to concerns about possible abuse of its lending powers for local "pork barrel" projects (Barnekov, Boyle, & Rich, 1989, p. 72). After the rejection of these proposals for an urban development bank, it became apparent that the Urban Development Action Grant program would, on a smaller scale, serve as the administration's basic instrument for leveraging private investment for local economic development.

UDAG and Urban Economic Development

UDAG was established in 1977 in the wake of an economic recession that had particularly serious effects in economically distressed cities. Created early in the Carter administration, UDAG was an attempt to respond particularly to the needs of such cities by making discretionary grants to them for projects most likely to attract new private investment. Eligible cities could submit proposals to HUD for projects that would involve both local government and private sector participation, with the proviso that those projects had to be sufficiently promising to lead to private investments that otherwise would not have been made.

The typical range of investments included office buildings, shopping malls, hotels, industrial parks, and upscale residential projects. UDAG provided loans and interest subsidies for private investors, as well as seed money for local government acquisition and clearance of land in redevelopment areas and for construction of related public infrastructure such as access roads, sidewalks, street lighting, and other utilities. By 1985 it had granted $4 billion to 2,700 projects in over 2,000 cities. Overall, UDAG served as an updated version of the earlier urban renewal program, but this time more closely targeted to encouraging private investment specifically for the revitalization of economically declining cities.

Like urban renewal, the focus of UDAG was on physical redevelopment as a means to local economic growth. But unlike urban renewal, UDAG targeted distressed cities and included explicit requirements for job creation for unemployed and low-income persons (Jacobs & Roistacher, 1980). In addition, UDAG tended to publicly formalize and regularize the interinstitutional relationship involved in the government-business partnerships that underlay the urban renewal program. In that program, using federal grant subsidies the local renewal agency purchased and cleared renewal tracts to be sold at a discount (write-down); then it would cast about for private entrepreneurs willing to enter into competitive bidding for development parcels, without any certainty that a suitable developer would be found. Thus in many cases renewal-cleared parcels of land stood undeveloped for years as vacant eyesores bringing no new revenue to the city because developers were unready to take on the risks of inner-city redevelopment.

By contrast, in the UDAG program business entrepreneurs were from the beginning brought into a bargaining process with city government officials to produce a mutually acceptable arrangement for a development project. Unlike urban renewal, in addition to indirect government aid such as provision of public facilities the development firm could receive a direct cash subsidy in return for its agreement to invest a specified amount in the project. Once an agreement was struck, an application would be drawn up and submitted to HUD for a UDAG grant. This required that the city include specific estimates of the project's impact on its tax base, the number of jobs that would be created, and local government's capacity for undertaking the project, as well as measures of physical and economic distress that demonstrated the city's grant eligibility.

In addition, in a stipulation unique to the UDAG program, the development firm was required to enter into a legally binding commitment to invest in the agreed-on project. Yet despite demanding a specific contractual commitment, UDAG allowed a high degree of flexibility regarding the uses of its allocations—they could be applied to a range of purposes besides land acquisition, from provision of accompanying public infrastructure (such as streets and utilities) to assisting the private financing of redevelopment—hence it was not difficult to find investors who were ready to put up the required funding for project completion (Barnekov et al., 1989).

Another consequence of the UDAG program was that with federal assistance, local general governments were helped to sharpen their capacity for redevelopment project planning in interaction with private firms and investors. On the local (horizontal) level, UDAG helped create staffs of developmentally oriented city officials with the skills to attract and fully service private sector investors, thus making it possible to move "beyond the single-project collaborations with developers that had characterized the urban renewal efforts of the 1950s" (Squires, 1989, p. 12). The result was to establish more complex, long-term public-private partnerships with access to an "array of subsidies, loans, and loan guarantees" that included but certainly were not limited to UDAG (Levine, 1989, p. 22).

In some cities economic development planning and implementation came to be fielded out to specialized quasi-public corporations not limited by the legal or procedural regulations associated with official public sector agencies or officials. Although this could make for speedier project design and implementation, the resulting developmental "shadow governments" raised serious questions of public control over the considerable public resources involved. In general, the process of local program design and fiscal administration basically remained closed to the public, so that economic development was to a considerable degree privatized, a matter confined to private investors and local officials, particularly when interacting within the specially created organizational context of these quasi-public development corporations. Thus whatever economic advantages might accrue ultimately to the host city from the specialized access to public subsidies afforded by these development entities, basic issues of democratic control of the locality's public investments were obscured and neglected as development projects were shunted around city councils by secretive shadow arrangements for administering public-private development programs such as UDAG (Levine, 1989).

On the vertical axis, UDAG facilitated the creation of strong intergovernmental ties between major public decision makers such as federal program administrators and local development-oriented mayors and planning officials. Thus the program helped maintain and update federal-local relations in the service of physical redevelopment objectives that had been associated earlier with urban renewal. Concomitant with this, UDAG provided the first clear signs in the Carter administration of a shift away from programs targeted directly to the unemployed or the poor and toward business-oriented urban economic development.

Within that approach, UDAG grants subsidized local public investments aimed at leveraging larger matching private investments in commercial, industrial, or residential projects. The assumption was not only that carefully targeted public investments would encourage privately financed physical/economic revitalization and thereby expand the local tax base but also that the fiscal benefits would ultimately filter down to all city residents, including the poor. In practice, this new orientation contributed to the presence of a basic ambiguity in the Carter urban policy between the earlier direct antipoverty emphasis associated with aiming CETA programs at the hard-core jobless (as well as CDBG regulations on behalf of social targeting to lower income groups) and the UDAG program's relatively indirect strategy of public investment in private business development with its assumptions of trickle-down benefits for the poor.

By the late 1970s President Carter had embraced firmly the strategy of federally assisted local public-private partnerships as a cornerstone of his policy to help older industrial cities; his administration vigorously promoted these arrangements through the UDAG program (Frieden & Sagalyn, 1989). This strategy grew in importance as the decade drew to a close, only to be followed in the 1980s by an at least equally strong emphasis on conducting urban economic development through local public-private partnerships. However, despite this similarity there were significant differences both in the general tenor and the fiscal specifics of the Reagan administration's approach to urban development, which further complicated the issues of public control of such development and of its social-economic results—issues of continuing importance to all who are concerned about the quality of life in America's cities, particularly its major older cities.

NOTES

1. As Dommel et al. (1982) observed with regard to CDBG program implementation: "When decentralized decisionmaking is implemented to maximize local discretion and minimize federal controls, the outcome takes on the character of a crap game—you roll the dice and take your chance in the arena of local politics" (p. 242).

2. As of the late 1970s, a growing number of calls were made by urban analysts for "planned shrinkage" of population through withdrawal of services in depleted areas of older central cities (Breckenfeld, 1977; Starr, 1978; Thompson, 1977).

3. This diversion was continued under the Reagan and Bush administrations. In fact, compared to what it had been under Carter, there was a reduction from 20% to 15% in the portion of a local government's CDBG allocation that could be used for public services, a program function of particular relevance to low-income residents because it provides for services such as job training and day care. At the same time, low-income groups became more conscious of the functions of the program and how to gain access to them, partly due to the efforts of organizations such as the Coalition for Low-Income Community Development (CL-ICD, 1993), a coalition that indicates membership of "over 1,000 grassroots and national organizations interested in federally funded Community Development Block Grants . . . and economic development issues" (inside cover).

Since at least the early 1980s, the CDBG program has been a major source of community economic development funding; in this respect, with the termination of UDAG and of EDA economic development grants in the late 1980s, CDBG took on added importance. From 1982 through 1989 economic development spending averaged about 12% of annual CDBG entitlement expenditures, or a total of $2.2 billion over the period (U.S. DHUD, 1992, p. A-5). Further, as of the early 1990s, the percentage of CDBG funds formally required to go to low- and moderate-income communities increased from 60% to 70%, and there was hope that federal actors such as HUD and Congress might take steps to better ensure that "economic development funds which are intended to benefit low-income people will really do just that" (CL-ICD, 1993, p. 5).

With regard to the legislative/administrative history of efforts to improve the targeting of CDBG to low-income people, see the thorough review by Gramlich (1994) in *Targeting Times,* a publication of the Coalition for Low-Income Community Development. (This useful source on developments in the CDBG program is available by writing to CL-ICD, 513 N. Chapel Gate Lane, Baltimore, MD 21229.)

Locally, in the particular cases where low-income groups have mobilized and exerted sustained pressure it was observed that they have been able to win development funding from local government (CL-ICD, 1993, pp. 8-10), although it is not clear how widespread such gains have been nationally. Beyond the area of economic development, a 1992 study covering over 170 cities was reported to have found that cities could use effectively at least another $6 billion in CDBG funding annually "in order to meet growing social service needs and to address the physical deterioration of distressed neighborhoods" (U.S. Conference of Mayors, 1993, p. 12).

[8]

Federal Retrenchment
and the Future of Urban Policy

FISCAL RETRENCHMENT AND REAGAN'S NEW FEDERALISM

In sharp contrast to the Carter administration, the approach to urban issues under Ronald Reagan's presidency at no time was based on any attempt to define specifically an urban policy. Instead that approach derived from the president's general philosophy of government, namely, that intervention by government (particularly the federal government) in areas of economic or social affairs is disruptive to basic voluntaristic processes in American society—hence such intervention is to be avoided if at all possible. Accordingly, the perspective of the Reagan administration was that urban development was best left to the processes of the market and of social voluntarism. Urban problems, whether of fiscal stress, homelessness, or crime, among others, ultimately would be resolved by reducing the federal government's role in state and local affairs and by freeing the forces of the market to stimulate both national and local economic growth.

In effect, the Reagan administration had an essentially "nonurban urban policy." Its policy toward urban places and problems was largely indirect and implicit, determined to a great extent by the workings of nonurban policies that were considered of higher priority. Overall, the thrust of those policies was fully consistent with the administration's philosophy of drastic reduction of direct federal involvement with urban affairs, which came to be known as a "hands-off" policy toward cities. The main elements of the relevant policies (Cuciti, 1990, pp. 237-238) were as follows:

1. *Sharp reductions in federal taxes,* particularly through cuts in taxes on upper-bracket incomes, as well as tax credits to encourage investment and special allowances for depreciation of industrial and other capital. In line with the supply-side theory favored by the administration, cutting individual and corporate taxes was expected to liberate private capital for new investment and allow the economy to grow enough to increase significantly, rather than reduce, net public revenues from taxes.

2. *Major shifts in spending,* from domestic programs to defense programs. Despite statements that government spending would be reduced by economies in federal programs, it expanded rapidly, due in good part to unprecedented peacetime increases in defense expenditures. These increases, together with major tax reductions, produced strong pressures to cut urban and related social programs in order to limit the resulting growth in the federal budget deficit.

3. *Restructuring of the intergovernmental system,* by devolving federal responsibilities to the states and localities. State governments would play the major role as partners to the federal government at the subnational level. Direct relations between federal and local governments would be reduced to a minimum, and cities would have to rely on state governments to provide them some portion of federal grant funds.

4. *Reliance on private sector initiatives,* with an active role suggested for private business in relation to urban affairs. This included the establishment of public-private partnerships between business firms and local governments through which to promote urban redevelopment.

In regard to its aim of reducing the federal role in urban affairs, the Reagan administration pursued a two-pronged strategy. It sought to make changes in the federal grant-in-aid system by sharply cutting the volume of urban aid and by restructuring the intergovernmental program relations involved in its delivery.

The administration had its greatest successes with the first part of this agenda in reducing federal spending for urban-related programs. The basic approach was set by the Omnibus Budget Reconciliation Act (OBRA) of 1981 at the outset of Reagan's first term. This budget legislation reflected the objectives of the Reagan administration in undertaking deep retrenchment in numerous urban and social programs, although significantly expanding military spending and maintaining a minimal income safety net for the dependent poor. About $14 billion were cut from grants to localities between fiscal year 1982 and fiscal year 1984 and another $11 billion in grants to individuals, such as unemployment and food stamp payments. The Public Service Employment program was abolished by 1982, and funding was reduced for education, training and employment programs, mass transit, airport and sewer construction, and notably for the Urban Development Action Grant program. Overall, this represented a serious reduction in the funding of categorical grants that supported services provided by state and local governments; at the same time, increases amounting to $52 billion were authorized for military program expenditures over this period (Fossett, 1984, p. 154; Mollenkopf, 1983, p. 283).

With respect to restructuring intergovernmental grant programs, the major change achieved by the Reagan administration was the 1982 budget consolidation of 57 generally small categorical programs into nine block grants to state governments. These grants included such urban-relevant functions as health, education, and social services, as well as a grant for low-income energy assistance and one for community development in small cities. Overall for these functional areas the new block grants reduced program oversight by federal agencies and transferred basic administrative responsibility to the state level. Although the states welcomed this expansion of their administrative authority, the total budget authorization for the new grants was about $7.2 billion annually, a sharp reduction of approximately 25% from their aggregate funding as categorical grants. In addition, although the Urban Development Action Grant remained the major categorical program for distressed cities, its budget was cut by more than 30% (Harrigan, 1985, p. 389).

When President Nixon first proposed block grants during the early 1970s, his administration emphasized that its purpose was both to improve the delivery of intergovernmental aid for domestic needs and to reverse the Great Society's concentration of authority in the federal government by instead strengthening the role of state governments. In contrast, Reagan's version of the New Federalism carried with it, for the states, an explicit trade-off: Enhancement of state administrative authority was proposed again, but it now was offered only on condition of reduced federal funding. For the states and their congressional representatives, this was a considerably less attractive arrangement than what Nixon's New Federalism had offered. As a result, the Reagan administration was able to win passage of only a single sizeable package of grant consolidations in 1982, folding almost 60 categorical programs into nine block grants. A second smaller consolidation was achieved the following year when 17 manpower training titles of the CETA program were brought together in the Job Training Partnership Act (JTPA) of 1982 as a block grant to states. Later proposals for further consolidation of categorical grants in education and for consolidation of a number of welfare categoricals were rejected (Fossett, 1984, p. 155).

In the early 1970s one of the Nixon administration's main arguments for adoption of block grants had been that by providing greater flexibility in the planning and provision of services, they would serve as useful mechanisms for improving the administration of federal grants to states and localities. In contrast, Reagan's objective in the 1980s was to substitute block grants for categoricals, not essentially as a mechanism for rationalizing the administration of federal grants but as a strategic step toward cutting the federal grant system at its root. Underlying this was a long-held aim of eliminating the established Washington social service bureaucracy, whose power base was identified with the federal grant-in-aid system and its categorical constituencies in states and cities. In this perspective, the expanded use of block grants in place of categoricals would serve as a transitional stage to subsequent withdrawal of federal involvement, not only from the administration of domestic social programs but from their funding as well.

Reagan briefly foreshadowed this strategy in an address to the National Conference of State Legislatures: "The ultimate objective . . . is to use block grants . . . only as a bridge, leading to the day when you'll have not only the responsibility for the programs that properly belong at the State level, but you will have the tax sources now usurped by Washington returned to you" (Palmer, 1984, p. 49).

Early on, Reagan's attempts to consolidate categorical grants may have appeared to casual observers as essentially consistent with the initiatives previously introduced by Nixon's New Federalism. However, in his annual budget requests Reagan repeatedly showed little of Nixon's inclination to use funding increases in support of block grants, either as a means of gaining congressional support or of co-opting the Washington bureaucratic establishment. Indeed, "rather than rewarding block grants with favored budgetary treatment, as Nixon had done, Reagan recommended them for some of the deepest spending cuts in his entire budget" (Conlan, 1988, p. 161). Thus, although initially it may have been believed that he simply was following through on the earlier New Federalism program of consolidating categoricals into block grants, it soon became clear that he had a more ambitious agenda for changing that system. In short, Reagan did not aim simply at decategorizing the system; he ultimately sought to defund it. This is what underlay his efforts to shift program funding responsibility to state and local governments. Reagan's intention to undo earlier policies was not limited to the Great Society and its use of federal resources to support national urban policies such as the War on Poverty. It extended to the basic decision to eliminate federal funding of publicly provided social services, whatever the level—state or local—at which they were administered. His determination to do so involved a rupture not only with the policies of Johnson's Great Society but with those of Nixon's New Federalism as well.

WITHDRAWING FROM NATIONAL URBAN POLICY

Behind this important policy shift by the Reagan administration stood a theory of the emergence of a new American society after World War II and the impacts on its older big cities. In this theory of urban postindustrial transition were incorporated some distinctive assumptions about the relation of once-industrial cities to the changing national economy and to the role of government. The theory noted that over recent decades the United States had experienced a major set of changes, from an industrial economy geographically concentrated in northern central cities to an increasingly decentralized information-processing and service economy. Due to that transition, older large cities generally have lost their former status as industrial centers, with many experiencing only limited success in restructuring as administrative and service centers. Public policy makers reasonably cannot expect to change these basic structural developments; instead they are urged to accept these facts and find ways to adapt to them. In brief, the future of older cities cannot depend on action by government but only on those cities' ability to play a useful role in the larger economy, particularly that of their metropolitan area. From this perspective, such cities must consciously evaluate the evolving technological and market forces acting on them, assess their competitive strengths in that changing environment, and develop their own economic strategies accordingly (Savas, 1983).

In this vein the President's National Urban Policy Report of 1982 advised cities to improve their ability to compete with each other by, among other things, improving their business climate and their appeal for residents and tourists. Under the Reagan administration, there would be a shift away from national governmental intervention in urban affairs, such as the provision of federal grant programs to cities. With reduced dependence on federal grants, urban development would be left as much as possible to the operation of the "free market"—viewed as the best mechanism for expanding local employment and stimulating community renewal. Seen this way, it was logical for the federal government to withdraw from urban policy making and leave it to "individuals, firms and State and local governments" that "properly unfettered, will make better decisions than the federal government acting for them"[1] (U.S. DHUD, 1982, p. 57).

Henceforth, any governmental involvement in urban policy would be limited essentially to state and local governments. Thus, in the context of Reagan-era New Federalism, insofar as the federal government had a valid role to play it no longer would be with regard to providing specific urban programs.

Under the Reagan administration, the federal government's first priority would be the stimulation of economic growth (using supply-side strategies such as tax cuts for wealthy investors and tax abatements for corporations). In regard to states and cities, the administration maintained that under its version of federalism those subnational governments now finally would get the opportunity to manage their own affairs without federal interference.

However, in view of the impact of federal cutbacks on urban-related programs, this also meant the loss of sizeable resources with which to cope with the problems of cities, particularly those in decline. The only other institutional complex with economic resources that can compare with those of the federal government is private business, particularly the private corporate sector. Turning to the private sector to form development partnerships was not unprecedented; various cities had become engaged in public-private partnerships even before the Carter administration. The difference was that now these partnerships with private entrepreneurs would have to constitute the main outside source of support for urban economic development. In particular, older cities in decline could look no longer to federal programs as a significant additional source of assistance, even for federal grants that could help attract private sector participation in developmental partnerships. In such a free-market entrepreneurial urban policy, "there was no room for targeting federal resources to high cost, less productive, declining areas; instead, resources were to be directed to those areas the private sector had already determined to be productive and profitable" (Levine, 1983, p. 17).

Despite pleas from cities to the contrary, the Reagan administration was adamant that its first priority—stimulating economic growth—would be pursued best in this manner. Now declining older cities would have to find their own way to compete for whatever share they might capture of that growth by initiating, expanding, or intensifying economic development partnerships with the private business sector. The administration's rationale was that this would be more efficient economically than subsidizing such partnerships with federal funds. Accordingly, using only their own resources, cities now more actively than ever would have to find ways to attract investment by bringing the private sector directly into the local development process.

At the same time, under an elaborate trade-off arrangement first advanced by President Reagan in January 1982, it would have become the responsibility of city government, with whatever help it might get from the state, to take care of the welfare needs of its residents. This was a key feature of Reagan's 1982 State of the Union proposal to the states of an exchange of program funding responsibilities that offered a dramatic program trade-off or swap.

Responsibility for funding and administration of the $19.1 billion Medicaid program would have been taken over from the states by the federal government, which at the same time would have turned over to them all responsibility for the major welfare programs—Aid to Families With Dependent Children (AFDC) and food stamps (together costing $16.5 billion), with recipients heavily concentrated in large cities. Although this appeared to be simply a cost-saving trade-off, it was projected as the first step in a series of "turnbacks" or reassignments to the states of over 60 federally funded programs, to be completed by 1988. A special trust fund based on federal excise taxes was planned to help the states finance the reassigned programs from 1984 to 1987, after which the fund was to be phased out by 1991. The states then would have been given the choice of using the trust fund revenues themselves, either to continue supporting the reassigned social programs or for other activities (Palmer, 1984, p. 50). Although some states might have been able to maintain the various programs involved, many others lacked either the economic or political capacity to do so without federal aid. Before long, numerous state and local officials expressed serious opposition to this proposal, based on doubts that their jurisdictions would really gain from these arrangements, and the Reagan administration did not follow up subsequently with a legislative proposal to Congress.

In its budget for fiscal 1984 the administration proposed drastic funding cuts in public housing programs, in subsidies for mass transit, and in a variety of programs of direct relevance to urban low-income populations, including vocational education, nutrition, and energy assistance (Judd, 1988). The intention was clear—the objective was a virtually total divestment of federal support for public services of particular importance to older big cities. Consistent with its strong free enterprise philosophy, the administration's justification for such a cutoff of federal aid was that government social programs were themselves mainly responsible for the decline of urban initiative and the growth of urban dependency.

Proceeding from this perspective, successive administration proposals (beginning with the one in 1982) included the elimination of established federal programs relevant to urban needs (such as CDBG and UDAG) by passing them down (devolving) them to the states. (In the case of EDA's urban development programs, the administration proposed their elimination outright.) Owing to a major campaign by organized urban interests such as the U.S. Conference of Mayors in coalition with civil rights and social welfare organizations, the administration found a lack of response to its proposals in the Congress. Had the swap proposal been accepted by Congress, CDBG and

UDAG along with 40 other social policy programs would have devolved to the states in 1982. Although CDBG, UDAG, and EDA programs were retained at that time, the Reagan administration remained critical of them, particularly of UDAG and EDA, each of which had been implemented actively under the Carter administration to stimulate local business growth in economically distressed areas. By the end of the Reagan administration both UDAG and EDA's urban development programs had been eliminated, and only one new economic development program proposed: the *enterprise zone* program.

URBAN POLICY BY FISCAL INCENTIVES

Enterprise Zones and Urban Policy

I have noted that from the very outset the Reagan administration was highly critical of the major existing urban-relevant programs and persisted in attempts at devolving them to the states or eliminating them altogether. During the entire course of that administration only one new program was proposed for urban areas—a program for urban enterprise zones. That proposal dates from efforts of Republican Congressman Jack Kemp and Democrat Congressman Herman Garcia to bring business enterprise and job opportunities to depressed areas such as the South Bronx. After a few years of discussion, the Reagan administration produced a legislative proposal aimed at providing incentives to business to move into designated urban areas. Announced in March 1982 and revised over the next several years, it proposed creating 25 zones annually for a 3-year period. In specific, "businesses in these zones would have 75 percent or more of their corporate income tax forgiven, would pay no capital gains tax, and would pay no tariffs or duties in areas also designated by the federal government as 'free trade zones' " (Judd, 1988, p. 360). In addition, there would be relaxation of state and local environmental regulations and of regulations on occupational safety and health.

Reagan administration supply-side economic theory emphasized that producers should be assisted by the removal of government barriers to economic growth and that this would stimulate the creation of new jobs. In line with this, the enterprise zones concept offered a package that included not only regulatory waivers but also tax breaks as inducements to private firms to locate in areas they otherwise would not have chosen. However, the program for enterprise zones was not a comprehensive new urban policy, and few zones (only 75) were proposed nationally. In fact, the incentives offered to business to locate in the zones were relatively unimportant. The reduction of federal taxes would have only minimal effect; compared with prevailing wages in an area, taxes typically are not a major factor in business location decisions. Overall, the zones would not have generated additional business volume for the nation; they only would have tended to redistribute business from one location to another (Judd, 1988).

As noted earlier, throughout the Reagan administration its officials expressed concern that at public cost EDA and UDAG both tended to attract to distressed communities investments that otherwise would have been more profitably located and would have created more new jobs in other jurisdictions. As OMB Director David Stockman remarked, "EDA doesn't create jobs. It reallocates them" (Stanfield, 1981, p. 494). Nevertheless, in the case of enterprise zones the same officials supported zone legislation that appeared to do essentially the same. How can this be explained?

One important difference was that the enterprise zones program made every effort to eliminate the regulatory role of federal or other governmental administrators, who in the past normally would have tried to assure that businesses receiving federal assistance lived up to environmental and occupational rules and legal regulations. Now, however, not only would federal regulations be waived, but there also would be strong pressure on states and localities to set aside their own regulations as a precondition for federal funding for designated zones (despite numerous Reagan administration statements expressing strong support for local autonomy). Another, probably crucial, difference was that the program left the investment initiative essentially in the hands of private businesspeople, whose participation would be rewarded through tax breaks and deregulation, rather than "through conditional federal grant and loan programs which allow federal bureaucrats to review the investment and operating decisions of private sector officials" (Levine, 1983, p. 19).

Yet another important difference was that the eligibility requirements defined by the act were rather permissive and could allow priority to zone designations in cities based on their efforts to establish the most favorable business climate. Hence zone designations would not be concentrated necessarily among the nation's most distressed cities. Although each zone must meet HUD's UDAG criteria of "pervasive poverty, unemployment and general distress," such communities "could lie within otherwise healthy, fiscally non-distressed jurisdictions and regions" (Levine, 1983, p. 21; U.S. Senate, 1982). This left open the possibility of enterprise zone designations being made for poverty pockets of prosperous cities in economically expanding regions such as the Sun Belt as readily as for impoverished areas in declining northern industrial cities.

Even after repeated hearings many such issues seemed unresolved, and congressional skepticism ran high regarding the specifics of how enterprise zones would work. Questions included how many jobs actually would be created anew rather than simply relocated and whether small businesses were the best target for the program's efforts, considering their typically high rates of failure. There was also concern that this untested program would end up taking the place of proven programs such as UDAG, which were targeted more explicitly to investment in declining cities. Such concerns ultimately prevented congressional passage of the Reagan administration's proposed legislation. Nevertheless, by the mid-1980s the idea had caught on at the state level, such that 32 states had established 1,400 zones, many with the purpose of competing for federal support should enabling legislation finally pass the Congress (Barnekov et al., 1989, p. 121).

In general, at the state and local levels enterprise zones have been used since to complement, not replace, existing public programs. Like the Reagan administration's proposed zones, they emphasize the use of tax incentives to attract business investment. However, they make relatively little use of deregulatory measures regarding health, safety, or the environment, and they target whatever public resources are available from existing programs into the zones in order to support their implementation. Consequently, "far from creating an environment of 'no government' and 'unfettered free enterprise,' state enterprise zones typically are areas of concentrated public effort—in land use planning, infrastructure investment, public service improvement . . . business loan funds, business technical assistance, training of disadvantaged workers, and similar efforts" (Bendick & Rasmussen, 1986, p. 114).

Thus, in order to work at all, enterprise zones have required considerably more government involvement and expenditure than promised by their more enthusiastic advocates. Moreover, enterprise zones are more effective in providing tax benefits to large firms with substantial tax liabilities, rather than the small, new firms that were pointed to by advocates as major program targets but that rarely show large profits in their early years of operation. What those smaller firms need is "venture capital or other forms of start-up financing, a type of assistance not provided in the proposed federal Enterprise Zone program" (Barnekov et al., 1989, p. 122).

The Reagan administration's enterprise zone concept was an application of supply-side economics on a local basis. It was grounded in the notion that entrepreneurial capitalism, particularly adventurous small businesses freed of government regulations, could solve urban problems by investing in severely depressed neighborhoods, thereby creating new business activity and productive jobs. Instead the program they proposed actually would have been of most benefit to large, already established corporate firms setting up branches in these specially provided low-tax, low-regulation zones. Employers had the option of availing themselves of federal tax credits offered in return for hiring disadvantaged workers, but with no established quota there was little certainty as to what percentage of employees would consist of residents of the designated zone, as opposed to already existing employees or new hires commuting to the area.

To administration advocates of enterprise zones, the size of the firms entering the enterprise zones likely was to be of secondary concern because the overweening objective was to eliminate government involvement in city economies—whether in the form of government regulations on business or of federally planned urban programs—consistent with the claim that such involvement discouraged incentive and fostered dependency, hence was the essential cause of urban distress (Bendick & Rasmussen, 1986). Yet contrary to this view, there is considerable evidence that more than any other single factor it has been large-scale private disinvestment—industrial, financial, and other—that lies at the root of the most serious economic problems of older cities (Burchell et al., 1984). Despite this, the strongly antifederal line of thinking prevalent in the Reagan administration became part of the rationale for the dismantling of federally funded programs such as UDAG and EDA. The choice of enterprise zones as the preferred instrument of urban economic development followed from the administration's intention not to continue any federally funded programs for local economic development. Their preference was that any government stimulation of private investment should occur "through the less distorting national tax policy"—in short, by means other than federal grants and loans subject to public administrative discretion and review (Levine, 1983, p. 19).

Tax Incentives and Urban Disinvestment

The enterprise zones program would not be the first time tax policy was used for purposes of stimulating economic development in the post-World War II era. The irony was that enterprise zones would now be established to try to undo in a relatively small way some of the massive effects of earlier tax and related policies on central cities over the course of the past half century.

Federal tax subsidies for housing—allowing deduction of both mortgage interest and of property taxes from federal tax liabilities—long had favored the construction of new single-family housing in suburban areas. By the mid-1980s these deductions amounted to a loss by the U.S. Treasury equal to an equivalent amount of actual federal spending (i.e., a "tax expenditure") of over $35 billion. These tax policies simply reinforced the subsidies to new suburban housing that have been provided over the past 50 years by Federal Housing Administration and Veterans Administration loan guarantee policies. In addition, several types of tax allowances to business and industrial investment have tended to encourage investment in newer urban and suburban areas over investment in older central cities. In particular, investment tax credits (ITCs) reduce a firm's tax liabilities proportional to any new capital investments it makes in a given year; this provides incentives for investment in new business and industrial equipment. Accelerated depreciation allowances permit reduction of a firm's taxable income by more than the actual value of the physical depreciation of its plant structures and equipment in a given year (Luger, 1984).[2]

These major tax forgivenesses have tended to favor larger, capital-intensive, highly profitable firms (such as multilocational corporations) able to make substantial new annual investments over more labor-intensive, less profitable firms "that used older equipment (often in older cities)" (Glickman, 1984, p. 473). In fact the investment tax credit has been referred to ironically as an "urban disinvestment tax credit" (Smith, 1988, p. 55), whose widespread use has had devastating effects on northern central cities, leaving behind concentrations of unemployed labor as well as abandoned factories and workplaces.

These tax subsidies to new residential and industrial investment were part of the larger federal strategy of postwar Keynesian growth policies[3] intended to stimulate private investment and consumer demand. The objective was to prevent serious recessions and particularly the recurrence of major depressions such as that of the 1930s. The idea (derived in part from the work of the English economist Keynes) was that timely government intervention in the economy could prevent excessive upward or downward swings (seriously inflationary booms or recessionary busts), which in the American economy periodically had characterized the business cycle from the mid-1800s on, causing a general economic collapse in the Great Depression of the 1930s. Federal economic intervention to stimulate growth and prevent recession can take various forms, many of which essentially leave otherwise taxable capital in the hands of potential institutional and individual investors (particularly large corporate firms and wealthy individuals); typically this is accomplished through tax exemptions, tax deductions, and tax abatements on business or individual income or property (Kantor, 1988, p. 263).

With the memory of the Depression fresh in the minds of policy makers, the major emphasis in the immediate postwar generation was on preventing another serious downturn of the economy due to a lack of adequate investment or demand. Thus the federal government committed itself to a national policy of economic growth after World War II. Much of this progrowth policy has been directed to supporting the postwar process of mass suburbanization that provided the framework of expanded consumer demand for suburban housing, furnishings, landscaping, and related goods and services made attractively accessible by certain distinctive features of suburban community development such as mass commercial malls. The striking fact is that the impact of this policy was not simply economic, changing the way in which the economy worked but sociospatial, changing the basic forms of land use and community settlement in America. A key element of this process was the unprecedented exodus of millions of urban households that were encouraged to leave central cities thanks to a rich combination of loan guarantees and tax deductions, as well as federal investments in highway construction and in water and sewage grant programs that made possible not only residential suburbanization but also extensive suburban industrial development, constituting a major aspect of the deindustrialization of central cities.

In these processes of decentralization and deindustrialization, the importance of tax expenditures such as business investment tax credits and homeowner tax subsidies is that they have relatively low political visibility compared with welfare payments or other public social expenditures that typically are examined closely and discussed intensely in Congress and the media. As a result, they have been difficult to confront and criticize with much hope of public understanding or response. Since the ending of the War on Poverty by the early 1970s there has been a pronounced shift away from hotly debated social program expenditures—"social welfare"—to relatively quiet tax expenditures—in effect, "fiscal welfare" (Smith, 1988, p. 69).

Due to its low visibility, this shift was not discussed much until the 1980s. Thus its implications not only for the urban poor but also for much of the remaining urban middle class, as well as for the fiscal viability of older central cities, largely have escaped broad public scrutiny even now. As a result, serious issues of urban-relevant policy change were effectively depoliticized during a crucial period of transition. Not only were federal antipoverty policies on behalf of the urban poor abandoned, but there was a broad flight of private capital—industrial, financial, commercial, and residential—from older urban areas, negatively impacting on their economies, their tax bases, and their capacity to provide public services for working- and middle-class households

(Burchell et al., 1984). This flight of capital from older industrial cities was encouraged by tax incentive programs at all levels of government, ranging from federal tax breaks for corporations and suburban homeowners to state and local tax exemption and abatement programs that tried to lure industry to the numerous state, county, or city jurisdictions advertising these tax breaks.

Tax concessions alone do not determine business location decisions because the most important factors in these decisions are labor costs, skills, and unionization, as well as transportation costs of access to markets and suppliers. But tax concessions reinforce these market factors (Glickman, 1984), particularly for relatively "footloose" routine-manufacturing firms or branch plants of northern corporations (e.g., in textiles or electronics) that tend to pursue low-wage, low-unionized locations, often in the South (Smith, 1988, p. 103). This pursuit most often shifts jobs from one location to another, rather than producing new jobs. With regard to the largest firms, "although 80% of the new business investment and depreciation write-offs promoted by Reaganomics have gone to America's 1,700 largest corporations, in the past two decades these employers have accounted for only 4% of new employment. Yet $750 billion in 'fiscal welfare' has been given up by the federal treasury as a result of such cuts" (Smith & Judd, 1984, p. 189). Not justified in terms of net jobs gained, these uses of tax expenditures as incentives to business investment represent massive subsidy to private disinvestment especially from older northern central cities, significantly contributing to their economic and demographic decline over the past two decades.

The preceding discussion suggests that adequate understanding of the decline of older central cities requires an awareness of the significance of private disinvestment and its relation to tax incentive/tax expenditure policies, which today constitute a massive but little recognized system of fiscal welfare for large capital-holding and investing interests. In an extensive study of urban distress and the resulting intergovernmental dependence of declining cities, Burchell et al. (1984) found that "many of the problems currently facing the intergovernmental city stem, primarily, from tremendous needs in private investment in its economic base. Put quite simply, private investors abandoned the older core city for more lucrative endeavors elsewhere" (p. 1).

Private disinvestment in older central cities has resulted in widespread plant closures, business bankruptcies, rising unemployment, property abandonment, and nonpayment of taxes. Although disinvestment has taken several forms, among the most important has been industrial disinvestment, largely through reductions in local investment in new plant and equipment, simultaneous with or followed by new investment elsewhere—in the suburbs, in the Southern Rim, or outside the continental United States altogether. Over recent decades, continuing industrial disinvestment has led to large-scale shifts of manufacturing jobs out of central cities, often followed by various forms of commercial disinvestment involving the migration of retail business and corporate and professional services to suburban malls and office complexes (Solomon, 1980).

Financial disinvestment further compounds the problems of older cities, often occurring when private lenders refrain from investing in areas with high rates of bankruptcy and foreclosure in the wake of industrial and commercial disinvestment. Withdrawal of sources of financing reduces the supply of venture capital for new business start-ups and of capital for expansion of already established business firms. This leads to a self-reinforcing spiral of disinvestment and to the decline of the city's credit rating, so that local government finds itself cut off from ready access to the private market in municipal bonds with which to finance needed improvements in public infrastructure, from roads and bridges to sewers and water mains. In the worst case, the city's basic economic efficiency is affected, its tax revenues decline, and it becomes strongly dependent on intergovernmental aid to meet rising demands for social services from the low-wage, irregularly employed segment of the local workforce. If external aid fails to keep pace with these escalating demands, severe mismatch between the city's revenues and expenditure may develop, eventually leading to significant reductions in public services and a further decline in the quality of city life.

Private disinvestment and the flight of capital from distressed cities can be related in part to the considerable power of multilocational corporations to shift plant investments away from older urban industrial locations. As a result, urban economic development officials have become especially concerned regarding the investment decisions of large national and multinational corporations. Over the past two decades cities have been striving vigorously to attract new firms, particularly large ones, and induce them to make investments in land and operating facilities within their own boundaries rather than some alternate location. Much of this effort may be in vain because large firms for some time have been moving out of the country altogether to employ cheap foreign labor and take advantage of the opportunities to avoid tax payments that are offered by many industrially underdeveloped countries. Nevertheless, it has become quite common for urban communities to engage in "bitter competition over capital investment decisions of large firms" (Kantor, 1988, p. 170). This has pitted cities against each other for the jobs and taxable income that corporate investment can bring to a community,[4] typically leaving large corporations in a favored position "in an investment process that gives them powerful bargaining advantages in determining the course of urban development" (p. 170). In general, the greatly increased postwar mobility of capital has become a significant challenge to the capacity of urban localities to implement sustained economic development. As a consequence, declining cities in particular have felt strong pressure to give highest priority to a variety of programs for economic development and considerably lower priority to investing resources in redistributive social programs (Kantor, 1991).

Unfortunately, instead of acting to counter these pressures, economic growth policies of the federal government have often reinforced intercity rivalry for capital investment, rather than containing and mitigating its effects. In the early post-World War II period, it was publicly visible expenditure programs (direct public investments) that were the rule—first to pay for infrastructure in support of suburban development and then for urban renewal infrastructure in support of CBD redevelopment, necessitated by a combination of earlier neglect and increasing suburban competition. However, at least since the early 1970s it has been low-visibility federal tax expenditures, especially in the form of various business tax breaks, that have played a predominant role with regard to urban and suburban development. Between the mid-1960s and the mid-1980s, approximately .75 trillion dollars in tax breaks were distributed from the federal treasury (Piven & Cloward, 1982, p. 7). These governmentally created tax breaks have helped produce massive capital flight from older cities, all the while, however, giving the impression that this was simply the result of private market processes for which government has no responsibility (Smith, 1988).

In fact, though, these tax breaks are indirect public investments whose benefits have gone predominantly to large private capital interests with relatively little public discussion or critique. Thus significant revenues are each year kept from the federal treasury in support of the needs of powerful private interests, producing, in effect, a type of special private-interest revenue sharing in tune with the major theme of privatization advanced by the Reagan-Bush administrations. Contrary to the notion of the "public use of private interest" by means of tax incentives (Schultze, 1977, title), it has been powerful private interests that have come to use and dispose of massive public resources. For all the talk of the public sector leveraging private capital for public purposes, it well may be that the major leveraging has been of public capital for purposes of private profit.

THE REAGAN ECONOMIC
PROGRAM AND ITS URBAN IMPACTS

The particular contribution of the Reagan administration to these trends in disinvestment was to raise them to a new level of intensity, in part through a concentrated supply-side program of tax reduction on potential sources of private investment, which involved a combination of massive personal tax cuts and business tax exemptions. Making extensive use of these tax expenditures, the Reagan economic program underwrote geographic patterns of private investment nationally by effectively reinforcing the skewed distribution of growth to already growing cities and counties throughout the 1980s, with consequences for urban investment and disinvestment in places and people that likely will endure for some time. The basic legislative outlines of the Reagan program were established through two major congressional acts in the Reagan administration's first year in office; these acts dealt respectively with federal spending decisions (budget allocations) and with federal tax policies. One was the Omnibus Budget Reconciliation Act, the comprehensive spending bill passed by Congress in 1981 (discussed earlier). The other was the Economic Recovery and Tax Act (ERTA) of 1981, supplemented by the Tax Equity and Fiscal Responsibility Act (TEFRA) of 1982.

With regard to personal tax cuts, ERTA spelled out a reduction of almost one fourth (23%) in personal income taxes that mainly benefited upper-income households; at the same time, it specified "deep cuts in business taxation." In addition, schedules for accelerated depreciation of capital invest-ments "were simplified and shortened under the Accelerated Cost Recovery System (ACRS), and the Investment Tax Credit (ITC) was increased." As a result of these measures "effective corporate tax rates for nonresidential investment" were more than halved, from 39.6% to 17.8%; overall the tax cuts provided by ERTA for corporations and investors meant a long-term loss to the treasury of over $1 trillion, although a part of the cuts was later canceled by TEFRA (Glickman, 1984, p. 472). Personal and corporate tax cuts of such magnitude, together with massive increases in defense spending, made for rapid increase in the federal deficit. Thus began a series of significant in-creases in the deficit, which marked the Reagan era.

Going beyond the enthusiastic early rhetoric of the Reagan administration, what were the actual accomplishments of its economic program? Probably the single major accomplishment (owing largely to the 1981-1982 recession) was the reduction of inflation, which had been running at a rate over 10% in 1980, to 3% in 1983. However, this was achieved at great cost in full-time unemployment (11% in 1981-1982, the highest rate since the 1930s' Depression) and in lost output of approximately $1.2 trillion, or an average of $15,000 for every household in America. Contrary to campaign promises of substantial economic growth as well as reduction of federal deficits, what transpired was "a deep recession and annual deficits of more than $200 billion," which translated into significant reductions, rather than expansions, in national investment and savings (Glickman, 1984, p. 472).

The Reagan program had negative results for declining cities due in part to the recessionary effects of its economic policy, but also after the 1981-1982 recession due to the tendency of its tax incentive programs to direct growth to already existing growth areas and away from areas in decline. ACRS and the ITC both yielded greatest benefits to firms in growth areas. On the other hand, firms with poor earnings lacked enough taxable income to benefit much from the ACRS tax breaks; thus low-growth areas, characterized by concentrations of such weak firms, gained little from this program. In general, ACRS tended to favor "new capital goods over old, and . . . plants and offices in growing regions over those in older, declining ones" (Glickman, 1984, p. 474). Similarly, firms in high-growth areas were most likely to buy new equipment and thus claim investment tax credits, as well as gaining added depreciation benefits based on these new investments. In contrast to the targeting of incentives to distressed cities, the Reagan administration created a massive system of federal incentives that reinforced private investment in areas of growth—areas where it most likely would have gone anyway—and relentlessly pursued the defunding and finally the abolition of programs such as UDAG and EDA, which had offered declining areas some hope of assistance in attracting private investment.

Compounding these effects of the Reagan economic and tax policies were the administration's cuts in spending for social programs, which had particular impact on distressed cities. Thus for the period 1980-1986 total federal spending for nondistressed ("healthy") cities grew by 65.7% in current dollars but by only 15.3% for distressed cities. Although defense spending grew at almost the same rate in healthy and distressed cities (75.4% and 70.1% respectively), the disparity in percentage growth was tenfold for nondefense spending, with 60.4% increase for healthy cities versus only 6.0% for distressed cities, owing largely to the concentration of social program recipients in distressed older cities whose programs received deep cuts (Cuciti, 1990, p. 250). These current-dollar comparisons refer to disparities in spending increases.

However, taking inflation into account, real dollar spending for urban programs fell by 10.6% in fiscal year 1982 alone and by 8.1% in fiscal year 1983 (Glickman, 1984, p. 476).

The administration's rationale for these cuts was that such programs properly belonged at the state and local level and that if they were important to the people at those levels, they would be funded by them. In the next section I turn to those subnational levels of government to get a picture of the fiscal response of state and local governments to the retrenchment policies of the Reagan era and of related changes in intergovernmental relationships during that period.

THE REAGAN PROGRAM AND STATE RESPONSES

On assuming office in 1981, President Reagan brought with him a long-held objective to restructure the respective roles of national and state governments in the federal system. He claimed that his "dream" in this regard was to devolve significantly various aspects of federal authority to the states in order to bring about a "proper mix" of powers between the two levels of government. But this objective was tied closely to "the preeminent goal of the Reagan administration in the domestic public sector" (Nathan & Doolittle, 1987, p. 6), namely, to bring about a deep retrenchment in social programs at all levels of government. For underlying the aim of increasing states' authority was the assumption that given the right signals from the federal level, the states would follow suit in reducing funding for public social programs—in short, that retrenchment would accompany devolution.

In case studies conducted in a sample of 14 states and their major cities during Reagan's first term, Nathan, Doolittle, and a team of research associates (Nathan & Doolittle, 1987) investigated state (and selected local) government responses to the 1981 federal budget cuts in grant-in-aid programs. The study examined these responses against the background of changes in federal aid in real dollars during three successive periods: (a) the last 3 years of the Carter administration, 1978-1981, when a combination of the aftermath of recession and the tax revolt that began in California in 1978 forced spending reductions in federal grants; (b) the Reagan cuts in fiscal year 1981; and (c) fiscal years 1982-1984, when the cuts stopped. Changing federal outlays and state responses were traced in three basic grant categories: entitlement grants (passed through states as payments to persons eligible on the basis of income, age, and other selected social variables); operating grants (for a wide variety of health, education, employment, training, and other social programs, as well as community physical and economic development programs); and capital grants (for highway and other public work construction and major equipment purchases).

During the Carter administration, the years 1978-1981 mark the beginnings of a significant decline in real (inflation-adjusted) federal spending that was to continue until 1982. Although total federal outlays over this period grew in current dollars, they actually fell by 6% due to rapid inflation. In real terms, operating programs were cut most severely, by 18%, and capital grant programs declined by 7%. These large downward trends in federal spending coincide with the emergence of a conservative tax revolt in various states (ushered in by California's adoption of Proposition 13 in 1978), which served as an important restraint on Carter's earlier plans for a national urban policy. Yet despite the reductions in operating and capital grants, entitlement programs actually grew significantly, by 16%, with notably large increases in Medicaid, food stamps, and public assistance. This attempt to protect economically disadvantaged entitlement constituencies (of whom many were concentrated in older cities) ended with the Carter administration (Nathan & Doolittle, 1987, p. 50).

Encouraged by the growth of the tax revolt movement, the Reagan administration used the Omnibus Budget Reconciliation Act of 1981 to make major cuts in federal grants-in-aid. OBRA cut federal grant outlays in fiscal 1982 by 12% in real terms, with the largest percentage reductions involving operating grants closely relevant to urban needs, such as education, training, employment, and social services. These reductions altogether entailed a cut of $5.6 billion for an overall reduction of 17% in this category. Capital grants for public infrastructure also took a sharp cut of 12%. On the other hand, entitlements (which include programs such as AFDC, Medicaid, food stamps, and child nutrition) were reduced overall by only 5%, which on the face of it, seems a comparatively moderate cut. However, this reduction came amid the most serious recession since the 1930s, which caused major increases in unemployment and thus significantly expanded the number of persons eligible for such programs. Normally, sizable increases rather than decreases in spending for programs such as these would be expected. But OBRA-based cuts were sufficiently large to reduce program spending considerably below what it would have been otherwise in a recessionary period. In the case of AFDC, for example, income standards determining eligibility were tightened to a point where caseloads in half the states in the sample dropped far enough in 1982 that program expenditures also dropped despite the deep recession in that year. In particular, working AFDC parents became a "fast-vanishing category," falling from nearly 20% of the total AFDC caseload to not much over 5% in several states sampled (Nathan & Doolittle, 1987, pp. 70-71).

State Responses by Grant Types

Overall, the "1981 budget cuts were concentrated in two areas: entitlement programs that assist the poor and operating grants to state and local governments for a wide variety of programs" (Nathan & Doolittle, 1987, p. 52). Of the two areas, federal cuts fell most heavily on entitlements targeted to the poor and particularly to the working poor.

How did the states respond to the 1981 cuts? Generally, with the exception of the more liberal states, state responses to cuts in entitlement grants other than Medicaid essentially were to take no compensating action and to pass them on directly to the affected recipients. Why the difference in response to Medicaid? Examination of the different grant constituencies involved helps to understand why states responded differently to cuts in that program. Like AFDC, food stamps, and the school lunch program, the Medicaid program's constituency consists of financially needy recipients. However, over half of Medicaid benefits go to elderly persons, typically a politically active group with relatively high levels of voting turnout. In addition, other important constituencies of Medicaid consist of powerful service providers such as hospitals and physicians. Overall, states tended to provide replacements for funding cuts in programs such as Medicaid that had politically influential constituencies; funding replacements came either directly from own-source revenues or through other devices, such as carryover funds or fund shifts from other programs. With its broad and influential constituency base, which often also included the families of elderly recipients, Medicaid was indeed exceptional. More typical, however, "as a general proposition, we found that the stronger the redistributive purpose of a given grant program (i.e., redistributive to the poor), the less likely was it to be protected by state and local governments from the effects of cuts made in federal aid" (Nathan & Doolittle, 1987, p. 96).

In the area of operating grants, state responses were structured in part in relation to Reagan administration initiatives that created or modified nine block grants that consolidated 54 existing categorical programs. These grants included the Community Development Block Grant, a Community Services Block Grant (CSBG), several health block grants, and block grants for social services and for elementary and secondary education. The CDBG was revised to give states the option of taking over the small-cities portion for cities under

50,000 population. This meant creating a small-cities block grant program under CDBG and allowing the states basic authority over allocation to these cities of program funds, consistent with the Reagan objective of strengthening the states vis-à-vis Washington in the federal system.

Although OBRA cuts in CDBG amounted to 6% for fiscal year 1982, Congress temporarily made up for these cuts with the Emergency Employment Act of 1983, which provided an additional $775 million to entitlement cities (over 50,000 population) and $223 million more to small cities. In long-term perspective, small cities fared relatively well under the new program because their share of the total CDBG allocation was increased from 25% to 30%, but this meant that an increased number of entitlement cities identified by the 1980 census now would have only 70% of total CDBG funds allotted among them (Nathan & Doolittle, 1987, p. 81). As a result, the CDBG small-cities program became the focus of sharp conflict between the administration and local officials in larger cities, who suspected the increase in its allocation was "the first step toward elimination of direct federal-local funding of the entitlement jurisdictions" (Farber, 1989, p. 29).

Also of special interest were state responses regarding the Community Services Block Grant, which funds community-based programs in health, housing, employment, and other social services for low-income persons. The funding recipients at the local level are community action agencies, created under the 1965 Economic Opportunity Act and funded directly first by the Office of Economic Opportunity and later by the Community Services Agency, both federal agencies. The 1981 budget act changed both the administrative relationships and the level of funding for the program. Administrative oversight shifted from federal to state government. Now the state would receive federal funds for this block grant with authority to decide its distribution among recipient communities. With regard to funding, the percentage cut in total CSBG funding was the largest among all the block grants (34%). However, under the revised administrative arrangements, no state replaced the federal cuts from its own revenues, and neither did any local government. Moreover, with their new authority to allocate funding, a number of states spread grant funds away from big-city community action agencies to agencies in other communities, so that less federal funding was distributed among a larger number of jurisdictions. In fact, the impact of cutbacks in spending associated with the 1981 block grants was felt most severely in large cities, not only in community services but also in the elementary and secondary education block grant because of these types of distributive allocation decisions by state governments (Nathan & Doolittle, 1987, pp. 79-80, 86).

Among the operating grants, the largest federal cuts were imposed on the Comprehensive Employment and Training Act block grant. CETA had included a public service employment program of considerable importance to maintaining public employment and service levels in large cities; this program was terminated. The training programs that made up what remained of CETA were replaced in 1983 by the Job Training Partnership Act, a new block grant limited to training only. Under CETA, federal funding for the most part went directly to localities, namely, to cities and counties over 100,000, as prime sponsors that acted to administer the program. Under JTPA, program funding goes to states, and major administrative responsibilities are delegated to councils composed of private industry executives in service delivery areas (generally large cities or consortia of smaller local governments). Emphasis is on training for private sector employment; no funds are provided for creating public jobs. Generally, training is short-term and on-the-job, hence more useful to persons with prior work experience, rather than the longer term classroom training required to impart basic skills to the hard-core unemployed.

Termination of the public service employment (PSE) component of CETA together with cuts in expenditures for job training accounted for spending reductions of $3.4 billion in current dollars, over half the 1981 OBRA cuts in federal grants. There was little or no attempt by states or local governments to replace these cuts with funding from their own revenues or to cope with them by using carryover funds or shifting funds from other programs. Overall, the most serious impact reported due to the 1981 cuts was experienced by low-income persons who no longer could receive income from the terminated CETA-PSE program (Nathan & Doolittle, 1987, pp. 86-87).

With regard to cuts in federal grants for capital programs, the responses of states generally were quite protective of the major programs (with one notable exception). In highway construction the 1981 federal cuts were small and were followed by substantial funding increases in the next 2 years. Sampled states generally raised their spending on highway construction and maintenance over this period, in part due to the "perception that road, bridge, and public transit facilities were essential to promote economic development" (Nathan & Doolittle, 1987, p. 90). Similarly, states and localities increased mass transit spending in response to anticipated as well as actual cuts in operating subsidies included in the federal grant for mass transit capital construction. They also undertook various coping measures to access alternative funds for wastewater treatment construction in response to 1981 budget rescissions and possible further federal cuts in that program. However, in response to sizable cuts in a number of federal grants to public housing, states and localities made little effort to provide replacement funding. This was consistent with the lack of compensatory effort in other programs (e.g., entitlement programs such as AFDC and food stamps and operating programs such as community services and elementary-secondary education) that had significant redistributive functions, generally of considerable importance to urban poor populations (Nathan & Doolittle, 1987).

Overall, review of the findings of this important research effort reveals a consistent pattern of state response: protection of funding from actual or anticipated federal cuts for capital programs that are developmental in nature (such as highway construction and mass transit) while passing on cuts in entitlement or operating programs that are redistributive (such as AFDC and food stamp entitlements and the community services and CETA public employment programs, respectively). In the case of the already distributive CDBG, further spreading of allocations away from large cities made it even more a distributive program, although regulatory changes allowing for greater emphasis on economic and physical development enabled shifts in program orientation within communities that further reduced its moderately redistributive effects.

LOCAL RESPONSES TO THE REAGAN PROGRAM

In a related paper based on information from field associates of the preceeding study, Wood and Klimkowsky (1985) examined closely the response behavior of cities to state actions in a sample of eight states. In their perspective, the Reagan version of New Federalism created for cities a "new ball game: more complex, more threatening, more uncertain than at any time since World War II" (pp. 229-230). For American cities, by the mid-1980s circumstances had changed for the worse, even compared to the period of big-city fiscal crisis in the late 1970s. Instead of temporary, localized fiscal crisis, the prospect for cities—in a context of growing national budgetary deficits—had become one of continuing, significant cuts in urban-relevant federal programs with little hope for improvement in the near future. Moreover, the Reagan budget cutbacks would be applied now as part of a policy of federal withdrawal from responsibility for the condition of cities and the vesting through block-grant federalism of "new power and discretion" in the states with regard to their localities. As Wood and Klimkowsky note, this devolution of power from the federal to the state governments is the "structural change that most sharply distinguishes urban fiscal experience in the eighties" (p. 230). (Further page references in this section are to the Wood/Klimkowsky study, except where noted otherwise.)

Overall, these analysts hypothesize that "the emerging intergovernmental system for authoritatively allocating resources to localities" would be "more unstable than its predecessor" (p. 231), due in large part to its greater complexity.[5] The preceding period of Cooperative Federalism from the 1930s to the mid-1970s was characterized by fixed patterns of influence relations ("iron triangles") between entrepreneurial mayors, responsive liberal congressmen, and supportive federal program managers, and by well-established urban interest coalitions, such as USCM and NLC. By contrast, state-local influence patterns and interest relationships under the latest New Federalism are considerably less systematic or predictable, so that the local game of pursuing state aid can be said to involve many more wild cards, such as "loose networks" of influence relations, and "shifting coalitions" of sometimes cooperating and often conflicting interest groups and interested jurisdictions (cities, towns, counties) concerned with urban-relevant policies and programs—all of which is further complicated by the rise of minority group leadership in urban politics and the uncertainties of state response to these new leaders. In this fluctuating interorganizational environment, cities will seek first to "maximize their own resources" on the basis of their particular socioeconomic needs and political-governmental structures. Although this likely will produce a variety of city patterns of approach and accommodation with state leaders, the overall effect of the looseness and uncertainty of these new intergovernmental influence-relations is "a diminution of city capacity to secure resources and authority from outside the local system," particularly regarding programs to meet the needs of the poor, which probably will be hit hardest by the intergovernmental politics of fiscal scarcity (p. 232).

In general, regardless of differences in political culture and economic and social characteristics of the cities in the sample, there was an interesting uniformity in their responses to federal program cutbacks. Where they had the authority, cities raised property tax rates; they also ventured to institute new taxes (such as hotel, motel, payroll, and income taxes) but often were stymied by state restrictions or local charter limitations. Thus they also turned to raising fees or establishing new ones for a variety of services, including water and sewer connections, health services, and trash collection. Some cities sold off assets such as convention centers or school property to their states or to private parties. Cities generally took steps to reduce expenditures especially with regard to costs of personnel through wage freezes and cuts, reductions in allowable overtime, and layoffs. Cutbacks in services included reductions in so-called soft programs such as job training and nutrition and even in undeniably basic services such as police, fire, and health (pp. 242-243).

Usually, these measures tended to magnify the impact of the OBRA cuts because their effects fell most heavily on populations already hit by cutbacks in federally financed employment and training programs, AFDC, and food stamps. However, as Wood and Klimkowsky (1985) put it, because the consequences of these various budget cuts "were not immediately visible, mortgaging the service future in people terms was politically acceptable" (p. 243). Similarly, the future was mortgaged also in regard to the physical infrastructure of the cities; various capital projects already underway were halted and deferred maintenance became the practice for existing public facilities. In addition, some new (and some traditional) management techniques, including office automation (and agency reorganization) were adopted widely among cities, various of which established central management systems, cost-benefit analysis, and system forecasting, as well as new techniques in controlling funds, materials, telephone use, and employment of contractors.

Fiscally speaking, with regard to pressures on cities produced by federal cutbacks, city responses were piecemeal and incremental (p. 243). To the extent that there was an underlying fiscal strategy, it was to use the least politically offensive measures, whether in the form of new taxes and fees to increase revenues or of cuts in payrolls and programs (including deferment of capital outlays) to reduce expenditures. Because these actions were often narrow gauged, they were too small individually to stir much opposition but insufficient in the aggregate to fully resolve the fiscal problem.

With respect to intergovernmental politics, however, there was one interesting constant; sampled cities generally tended to focus on making the most of their own resources rather than asking for large-scale help from their states or entering into agreements that might undercut their local powers. Seeking to preserve "whatever power and autonomy they possessed" they did not buy into the program of devolution "which Reagan had urged on the states." Hence these cities "did not ask of the states a comparable increase in local authority and in local resources" (p. 244). Clearly, big cities are not natural constituencies for elected state officials, who are more likely to look to suburban voters for support; hence the cities rate low in political priority with them. This is perceived fully by city officials, who expect that in turning to their states for fiscal aid, they are more likely to receive expensive mandates than additional moneys (p. 246).[6] In addition to worrying about mandates, city officials have reason to be concerned about serious problems of displacement of federal aid to state rather than local purposes under New Federalism. Thus Wood and Klimkowsky quoted Luce and Pack (1984) who, based on their own study of state behavior toward localities, observed that "in many states, even (federal) aid cutbacks of 50 percent are not as harmful to local revenues as the displacement of federal aid which occurs *when the state is made the intermediary between the federal and local governments*" (p. 354).

In short, as part of its experiment in devolving federal responsibilities to lower levels of government, the Reagan administration sharply reduced federal programs of aid to states and localities. Cities typified by high levels of demand for public services and poor fiscal resources were most seriously affected. In such cities, the impact of program cuts fell most heavily on the poor and their children and on other economically vulnerable groups, such as the elderly, the handicapped, and the chronically ill (Wood & Klimkowsky, 1985, p. 248). Frequently, the response of such cities was an improvised potpourri of tax increases and local program and payroll cutbacks; insofar as a broader strategy could be identified, it was that of postponing the full impact of federal cuts while waiting for the economy to revive and for local management improvements to make themselves felt.

Unfortunately, for all the improvements in their administrative and fiscal capacities in recent decades (DeGrove & Brumback, 1985) the response of state governments to the distress of seriously impacted cities was quite minimal. States were reluctant to fill the gap in financing or providing services in the wake of federal withdrawal; they tended to limit themselves to partial replacements of basic services. In their own areas of discretion, they were not ready either to pass federal grant money through to distressed urban areas or to reduce significantly their restrictions on local authority to raise revenues (Wood & Klimkowsky, 1985, p. 248). In addition, federal cutbacks left it to the states to play the role of funders for local programs; this meant that localities now would have to disperse much of their lobbying efforts from Washington and lobby separately in 50 different state capitals on behalf of funding for programs affected by cuts. Cities would have to construct anew at the state level interest networks with state legislatures and program agencies to replace the Washington networks that they had built up over years of patient effort.

Under the New Federalism, the intergovernmental city does not disappear. Rather, it shifts its dependence on external assistance from the federal government to the level of the state governments and their agencies—a level at which cities find themselves relatively disadvantaged, in part due to the absence of comparable influence networks to those previously established in Washington and in part because of the higher political priority accorded to increasingly powerful suburban interests in the states. In the final years of the 20th century, despite professionalization and modernization of state governments, state politics and politicians generally are more attentive to matters of suburban growth and development than to problems of big city poverty and homelessness. As a result, Wood and Klimkowsky's (1985) view of the prospects of the cities in this new intergovernmental era was clearly pessimistic, and they saw "no detectable evidence that cities, even in time of a reviving economy, will benefit from the New Federalism" (p. 249).

LOCAL PUBLIC-PRIVATE PARTNERSHIPS: EXPLORING THE POSSIBILITIES

Based on the preceding discussion it is clear that until national government once again adopts a strong urban policy, cities must expect to proceed with relatively little help from higher levels of government. Instead they will have to turn to partners in the private sector to provide at least some of the resources local governments lack for successful urban redevelopment. In these circumstances cities will have to assess realistically the options open to them. The four models of public-private partnership that are presented here are by no means exhaustive. Instead they correspond to two extreme points on a spectrum of possible partnership types and two significant intermediate points; thus this typology allows for the placement of important existing subtypes or types that yet may emerge between these.

Unequal (Business-Dominated or "Corporatist") Partnerships

Unfortunately, local public-private partnerships all too often have been essentially unequal, "with the corporations doing the planning while the city government facilitates corporate plans using municipal legal powers" (Carnoy, Shearer, & Rumberger, 1983, pp. 197-198). As a result, the private partners receive the lion's share of the benefits beginning relatively soon after project completion; meanwhile, the city waits patiently for its tax payoff while providing various public services. Thus the downtown corporate center strategy associated with business-dominated partnerships generally has contributed to the emergence of a dual or bipolar city characterized by an increasingly two-tiered income distribution, with white-collar managerial, professional, and highly skilled technical types in the upper tier, marginalized blue- and white-collar workers in the lower, and a growing lack of adequate occupational ladders to bridge the gap between these tiers (not to mention the unemployed underclass who are largely outside the regular urban economy altogether). As Levine (1989) observed, "*Uneven* growth has been the logical outcome of *unequal* partnerships and closed decision-making processes" (pp. 27-28).

Preponderately unequal business-dominated partnerships constitute one extreme of the public-private partnership spectrum. At this end are the conventional unequal partnerships of the 1950s and 1960s associated with urban renewal, later succeeded by the updated unequal partnerships of the 1970s and 1980s. These partnerships have taken a number of forms, including conventional growth coalitions, quasi-public development corporations, and publicly formalized contractual relationships as, for example, in the UDAG program. What the partnerships of both periods have in common most significantly is the heavy dependence of local government on the participation and resources of a single partner-type (typically involving one or a few associated corporate business partners), the very active role played by that partner-type in the design and implementation of redevelopment,[7] and the relative imbalance of benefits to the respective partners (business and local government), as well as the relatively long time lag before benefits are received by the locality. Hence I view both the conventional and new kinds of unequal partnerships as subtypes of a broader business-dominated or corporatist redevelopment partnership type.

Linked Development Partnerships

Although local business elites have long dominated the orientation and content of redevelopment (particularly downtown redevelopment), their public partners still may be able to share more fully in the benefits provided by successful redevelopment programs. One strategy to make the public-private partnership less unequal (and at least more financially rewarding) involves linked development or linkage policy. Well-designed linkage policies can secure for the host city greater levels of benefit than it would receive otherwise (and perhaps reduce the level of negative impacts on city neighborhoods).

Linkage 1: Limited Linkage Partnership

In its more limited form, linkage partnership may be achieved through local government pursuing a calculated strategy of trade-off with its private sector partners. Local government here plays the role of a savvy public entrepreneur seeking to share in the profits of the joint development enterprise. It seeks to overcome the disadvantages of its lower access to relevant economic information and to development financing than its private partners by engaging in the urban development game as a shrewd self-aware poker player.

This means knowing "what financial roles [it] is capable and willing to play," and particularly resisting the temptation to play the role of gift-giver, lenient about giving "front end" concessions to its private partners. Instead it will hold out for the role of an equity partner that "can bargain concerning distribution of fees, net cash flow . . . and net proceeds." It also means tapping legal and accounting experts "knowledgeable about joint property ventures," then making a clear assessment of the risks of the venture and establishing a reasonable rate of return accordingly. In addition, the city will diversify its risks in investing public funds in the joint venture by tapping a broad range of proceeds including interest payments and preferential returns as well as fees and net proceeds. It further will limit its risks by arranging to "play with other people's money," using federal or state funding to whatever extent possible (Cummings, Koebel, & Whitt, 1989, pp. 219-220).

The possibility for linkage here is based on city government maximizing financial returns and avoiding financial risks through shrewdly negotiating the city's interest as a limited partner in the development scheme. There are only two types of players in this limited linkage game; no neighborhood representatives, union leaders, or small-business types participate. It is a guessing and outguessing game played between public and private elites seeking to gauge and exploit each other's informational and financial weaknesses or points of vulnerability. Because those vulnerabilities usually tend to be greater for cities than for the generally well-financed, well-connected, highly mobile developers they deal with, it is a partnership strategy that only occasionally can be very successful for city government.

The corporatist and limited linkage partnerships are differentiated for the most part only by the division of revenues associated with the joint development venture, generally focused on central business district redevelopment. In short, the limited linkage partnership is a type of local (i.e., horizontal rather than vertical) revenue sharing by the city's government out of the several types of proceeds that may flow from downtown corporate-centered redevelopment (rather than out of federally provided revenues, as in general revenue sharing). Like vertical revenue sharing, this leaves the city administration essentially full discretion to decide how to spend those revenues and thus determine the extent and nature of linkage. That includes whether, for example, to use them to meet administrative costs, to cut tax rates (benefiting middle-income taxpaying residents, who also are often active voters), or to provide added social services to the poor, who pay few taxes and shun the polls; thus it leaves the city broad discretion as to which constituencies and which neighborhoods to target for the resulting benefits of shrewd entrepreneurial deal making.

Linkage 2: Expanded Linkage Partnership

The expanded linkage partnership differs from the preceding in its attempt to distribute purposively the benefits of redevelopment more widely through the city's districts beyond the downtown, particularly to its lower income residential neighborhoods and their deteriorating commercial strips and public facilities throughout the inner city and the adjacent zone of gray areas. This relatively new type of development partnership (established in a number of cities including Boston and San Francisco during the 1980s) includes both "greater public control of the redevelopment process" and "more equitable distribution of benefits and burdens" involved (Levine, 1989, p. 29). In return for the rights to conduct redevelopment projects and for the public services and facilities that may be included in the deal, developers have been required to contribute to funds established to help meet the city's needs in critical areas such as housing, job training, public transportation, and a variety of social services and facilities (e.g., rental housing and a child care center in Santa Monica, California) (Shearer, 1989).

The emergence of expanded linkage partnerships has not occurred in a social-economic vacuum. They typically have been associated with the emergence of vigorous community organization and activism (often referred to as progressive or populist in character), the development in various cities of coalitions of such neighborhood organizations, and their entry into the urban electoral arena. In some cities this has included coalitions involving labor unions and community organizations; there has also been participation by local small business interests. This development is a phenomenon that thus far has been specific to several dozen older industrial cities (e.g., Detroit, Chicago, Cleveland, Baltimore), cities with large minority concentrations (e.g., Washington, San Antonio), and a sprinkling of university towns; but although specific to certain types of locales, it is spread broadly enough to be national in occurrence. However, lacking either a central organizational framework or nationally recognized leadership, it is not quite a national movement (Shearer, 1989). Clearly, its reform agenda and style are oppositional to the remaining old-line urban Democratic machines; at the same time, these present-day urban reformers are at least as much opposed to the bureaucratic, fiefdom-ridden governments that the early 20th-century reformers produced as they are to the remnants of the old machines.

From an economic perspective, linkage policies have worked best in cities with strong growing central-district economies engaged in large-scale redevelopment; the larger the downtown development boom and the longer it lasts, the better. Yet politically there has to be at least a basic level of recognition by civic leaders that large-scale redevelopment, however successful it may be from an investment standpoint, itself "causes or exacerbates" a variety of problems, including "displacement from gentrification, overloaded and underfinanced mass transit, unemployment among poor city residents, and neighborhood deterioration" (Keating, 1986, p. 134). Presumably, enlightened civic leaders, themselves often from business backgrounds, can bridge the gap between local political actors and corporate-development interests in support of programs to mitigate or ameliorate these negative impacts of redevelopment. But whatever the role played by such leaders, the major political variable in the local adoption of linkage policies has been the mobilization of the "support of progressive political slates and of candidates backed by neighborhood organizations" that oppose narrowly corporate-centered development policies (p. 134).

Thus the conditions for the formation of expanded linkage partnerships have included the emergence of antimachine political organization and leadership supported by a substantial degree of community organization and activism. Behind this lies the process begun in the 1960s of the broadening of urban politics by community organization and leadership development, traceable to the programs of the War on Poverty and the precedents they set in producing leaders independent of both the urban machines and the entrenched city bureaucracies. In short, compared to the limited linkage partnership and its narrow economic focus, emergence of the expanded linkage type has required addition of a political dimension along which the dynamics of community organization can be translated into effective political pressure on corporate development interests. The major objective has been to open development partnerships to other participants and to a broadened linkage agenda that extends at least to the pressing needs of minority and working-class neighborhoods, including many of the inner-city neighborhoods surrounding downtown.

However, expanded linkage per se is not necessarily the full answer to the needs of local economic development. For one thing, linkage programs have been ameliorative or compensatory; they have been used by local governments to ameliorate the deep reductions in urban-relevant federal aid since the early

1980s and to make up for constraints on local revenue raising imposed by state tax limitation legislation (Smith, 1989). In particular, they have been used to compensate for the negative effects of downtown development itself, such as gentrification of rental housing or crowding of the transit system. Thus linkage programs basically have not been developmental, although they might contribute to economic development through provision of funds for education or job training programs established to compensate for the relatively small number of jobs that corporate center development opens to low-income and minority workers (Keating, 1986).

Furthermore, in cities such as Boston and Hartford where linkage was intended to redistribute a portion of the proceeds of downtown growth to neighborhoods, the "required linkage payments" amounted to only "a tiny fraction of the total benefits flowing to downtown developers," who then simply passed the costs on to new tenants through "increased fees and rents" (Smith, 1989, p. 95). In recent years major developers have become more ready to accept linkage fees because they free them from becoming embroiled in struggles with community groups who bargain hard to increase linkage benefits. Commercial developers would rather pay a fee established by bureaucratic formula and avoid involvement in open-ended bargaining whose outcome is much less predictable. Hence, unless carefully managed, linkage may have the potential to become a type of "license fee" that allows the developer to proceed with projects that involve significant negative impacts on the community, having purchased a certain degree of immunity from effective political reaction or governmental prohibition (p. 97). Thus the interesting question raised by Smith, "Linked to what?" leads to further reflection. It helps point up the basic problem that although linkage programs depend on urban economic growth, they do not determine the nature or directions of development that produces that growth. Thus, used imprudently, linkage can become a license for unsound development or overdevelopment (as in the case of large-scale overbuilding of office space in various downtowns).

It is significant that Keating (1986), in an examination of the conditions required for successful expanded linkage policy, observed that the chances for adoption of linkage are increased "in cities that are developing new downtown plans or revising land use policies for their central business districts." As he further noted, "Such processes provide public forums for analysis of the costs and benefits of downtown development" (p. 134). The availability of a public context for analysis and debate regarding the full costs of the corporate center strategy and how its benefits may be shared more widely to at least ameliorate those costs is invaluable to the beginning of a broader process of planning—

planning that is concerned not only with amelioration through linked development but with the basic direction of development itself.

Strategic Progressive Planning

This model is at the opposite pole from the business-dominated public-private partnerships and quasi-public shadow governments with which this discussion began. The conditions for strategic progressive planning thus far have been difficult to meet on a sustained basis, so that this type of planning has not yet achieved the status of an institutionalized structure in city governance. Chicago in the mid-1980s provides an example of an intensive effort to establish such planning at the local level.

The April 1983 mayoral election of Harold Washington, a strongly reform-oriented or "progressive" Democrat and the first Black mayor in Chicago's history, "raised hopes that a major U.S. city might take significant new policy directions" (Shearer, 1989, p. 291). In particular, his election raised hopes for the establishment of a new planning agenda and indeed a new, more people-oriented approach to planning for the city's needs. Traditional city planning was concerned overwhelmingly with changing the built physical environment of the city. In successive stages, it moved from the objective of producing the City Orderly to the City Beautiful to the City Efficient. By the 1940s, city planning (particularly in large cities) had become master or long-range comprehensive planning. However, as Gans (1970) wrote, "Traditional comprehensive planning was (and is) neither comprehensive nor planning" (p. 241). Although it called itself comprehensive, this approach to planning focused on selected spatial aspects of the city as if it were little more than a collection of land uses and their economic values and gave minimal attention to the city's substantive social-economic problems. Its key activity consisted in the production of highly similar master plans, turned out mechanically in city after city (despite significant differences between those cities, socially, economically, and institutionally) and destined to gather dust on the shelves of planning agencies rather than serve as action guides to further city development.

In their long-range comprehensiveness these plans became grand catalogues of prescribed standardized land use, oriented to the long-term future with little relevance to the real, existing city and its changing social and economic conditions and needs. Typically they involved a long menu of standard planning prescriptions, such as the relatively detailed separation of land uses into different activity zones, the provision of numerous standard public facilities, the "creation of physically bounded and school-centered neighborhoods . . . the development of efficient central business districts and transportation systems . . . and the creation of an economic base which would maximize land values and tax receipts" (Gans, 1970, p. 241). Generally, the massiveness of the resulting documents of the master plan (which usually required at least several years to prepare) meant that any "focus on key elements is lost in the overburden of documents and maps . . . its complexity and bulk and the cumbersome procedures involved in its preparation rule out any simple process for modifying it or bringing it up to date" (Candeub, 1970, p. 218).

The 1984 strategic plan by the Chicago Works Together Planning Task Force—*Chicago Works Together (CWT)*—stands in sharp contrast to these traditional master plans. Here a specially appointed "development subcabinet" (composed of representatives from the mayor's office and relevant planning and development-concerned city departments) defined major issues and goals for Chicago's development and devised specific policies that would implement those goals. At the head of the list, Goal I was to increase job opportunities for the city's resident workforce, simply expressed in the slogan "Jobs for Chicagoans" (Judd & Ready, 1986, p. 230). This would be implemented by identifying "economic sectors that have promise for providing high-quality, lasting employment locally" (Mier, Moe, & Sherr, 1986, p. 308); then public investment would be directed into those sectors in support of employment development. In addition, the city would seek to buy goods and hire employees locally, rather than from external sources; this was intended to encourage the development of a "home-grown economy" rather than one dependent on external sources of labor and material (Shearer, 1989, p. 297). Associated with this were policies for investing in the development of a skilled labor force and affirmative action policies to provide expanded employment opportunities to minorities passed over in earlier planning for economic development (Judd & Ready, 1986).

Goals II and III had to do with orienting economic development planning toward the revitalization of (often long-neglected) neighborhoods. Goal II was to promote a better balance of economic growth between the neighborhoods and downtown. Goal III was to assist neighborhoods to develop economically through public-private partnerships involving neighborhood community development corporations (CDCs),[8] to foster neighborhood planning, and to expand neighborhood housing and local facilities through linked development arrangements.

Goals IV and V were of a political nature. Goal IV, to enhance public participation in decision making, reflected the strong interest of the Washington administration in opening up to all citizens the opportunity to participate in what had been a closed, exclusive style of city politics, dominated by elite decision making between machine bosses and business leaders of Chicago's central district (the Loop).

Like Goal IV, Goal V (to pursue a legislative agenda at the regional, state, and national levels) was also political, but it was projected along a rather different dimension of contemporary urban policy making. Goal V was somewhat unique in city planning not only for its explicit inclusion of political objectives but for its recognition of the reality and crucial significance of what I have referred to at various points as the vertical or intergovernmental dimension of urban policy in the late 20th century. (Viewed simply as an element of a document concerned with urban economic planning, the inclusion of such a goal might seem misplaced. However, there was a solid and even compelling political-economic background—the impact of Reaganist New Federalism, coupled with the impacts of the 1981-1982 recession—in whose context it is readily understandable.)

Even this brief summary of the 1984 Chicago plan for economic development differentiates it clearly as an action-oriented strategic plan from the long-range comprehensive plans that traditionally characterized city planning. The fact that it could be summed up in just five goals accompanied by 14 implementing policies itself marks the substantial difference from comprehensive plans, in which "10 or 15 goals and 50 to 100 policies and other recommendations have been common" (Mier et al., 1986, p. 308). Furthermore, the first three goals reflect clearly the nature of the 1984 plan as a strategic plan directed at using public powers of investment and of economic negotiation to change the economic structure of opportunity in the city of Chicago.

This is local government actively engaging itself in the restructuring of the industrial urban economy, this time along lines that will begin to take into account those who typically have been displaced by the conventional process of postindustrial urban restructuring, notably including entry-level, semi-skilled, and unemployed blue-collar workers, both White and minority. The particular approach to increasing economic opportunity that is adopted here suggests a process of strategic learning since the antipoverty experience of the 1960s. The result is an essentially structural strategy for reducing unemployment and poverty in Chicago, which includes expanding opportunities not only for education and job training but also most prominently for job creation, the key area of contention that the Johnson administration chose to exclude from its War on Poverty.

Nevertheless, despite its sensitivity to economic issues, the strategy incorporated in the 1984 plan is not simply economic. There is, as noted earlier, an explicitly political segment of the plan, which introduces (through Goal IV) the need to expand opportunities for nonelite citizens to be informed fully about and to participate in policy-making processes that affect their present and future quality of life ("life chances") and that of their families and local communities. The absence of such political opportunity means subjection to top-down planning of a not necessarily benevolent type. Thus Goal IV broadens the scope of participation and the nature of the planning process by including the dynamics of community organization and goal formulation in the process.

Goal V also changes the nature of the urban planning process by taking into account the fact that cities themselves are subject to top-down planning processes, in which decision elites at the federal and state level can give or take back public resources vital to city development and, in addition, may mandate responsibilities for public programs for which older cities in particular simply are unable to pay without either impairing their own developmental base or causing further hardship to their neediest residents. Goal V reflects the adoption of a consciously organized program to access whatever resources may be available for local development from other levels of the intergovernmental system and to lobby for legislation targeted to their special needs with whatever clout large older cities still can muster at the federal and state level.

As noted earlier, strategic progressive planning is premised on a special and not easily reproducible set of conditions occurring together. Besides a growing central district economy, it requires several political elements. These include not only the presence of progressive political organization (as in the expanded linkage partnerships) but also of a reform-oriented mayor concerned with creating a better balance of development between the city's haves and have-nots and notably between major downtown corporate/institutional interests and the less advantaged neighborhoods of the city. At the base of these elements there must be vigorous grassroots community activism and strong, competent community-based organizations willing and ready to challenge the developmental status quo.

The election of Harold Washington in 1983 embodied the emergence of these forces as a new coalition for balanced growth after many decades of dominance by the established progrowth coalition, composed of the old machine, downtown business interests, and segments of organized labor (such as public employee and construction unions). The new coalition combined reform-oriented middle-class White professionals, White ethnic dissidents, and a broad majority of the Black community, whose population by 1985 amounted to 41% of the city total (Preston, 1989, p. 119). But Washington's victory was by no means a complete capture of municipal government; instead, Chicago's government split between the still machine-ridden city council (the mayor eventually gaining only a fragile council majority) and the forces of reform in the executive branch. Even there, however, the implementation of Washington's programs was in danger of being paralyzed by a bureaucracy originally installed by the machine, which was often slow to carry out policy, acting (or failing to act) partly out of a mix of continuing political and racial bias and partly owing simply to the fact that it was "staffed by too many inadequately trained employees, including many planners in the development group" (Mier, 1993, p. 187).

As of February 1986, the development subcabinet made an assessment of the administration's performance with regard to the 85 programs and projects discussed in *Chicago Works Together*. (Certainly one of the subcabinet's achievements was to establish regular assessments by which its performance was subjected periodically to a systematic and public evaluation—probably something of an innovation in Chicago city planning.) Overall, the record showed 34% of program or project targets fully or more than fully achieved, 57% partly achieved, and 6% showing no progress. This was a reasonable record of achievement considering that only 18 months had elapsed since the publication of *CWT* in May 1984. It also was estimable considering the difficulties in bringing up to speed a slowly moving inherited bureaucracy to tackle the details involved in implementing programs such as the resurfacing of 5 miles of residential streets in each of the 50-odd wards of the city. This was a major component of the mayor's long-term program of refurbishing badly deteriorated neighborhood infrastructure, including "collapsing . . . sidewalks, streets full of potholes, clogged sewers, deteriorating alleys, undermaintained or nonexistent branch libraries, and overcrowded municipal facilities like health and social service centers" (Mier, 1993, p. 183).

The neighborhood infrastructure program not only was important in its material contribution to neighborhood life but also was symbolic of the mayor's determination to make up for the long neglect of neighborhood needs and interests associated with the downtown corporate-center strategy that prevailed from the 1950s through the 1970s. It also provided him with a way to reach out not only to minority areas that already were part of his basic constituency but also to long-undermaintained White ethnic neighborhoods. The intention was to fund the projects involved through a proposed bond issue for capital improvements sent to the council early in 1985. Faced with stalling tactics in the city council, Mayor Washington made a point of visiting the wards of key opposition aldermen in a bus filled with newspaper and television reporters in order to shift a handful of crucial votes. Walking the streets and talking personally to residents on camera, he was able to bring substantial popular pressure to bear on the council members whose wards he visited. This, coupled with his agreement to undertake complete street reconstruction in the key aldermanic wards at stake (which contributed significantly to increasing the proposed $50 million program to $160 million), made it possible to swing the vote in favor of the bond issue (Mier, 1993, p. 185).

Besides dramatically reaching out to White ethnic residents dissatisfied with neighborhood conditions, this allowed the Washington administration to connect with area small businesspersons, always an important local political force, as well as an economic sector that tended to employ local residents, rather than the suburbanites who often took the jobs offered by downtown corporate business establishments. In short, the neighborhood public works program was important to furthering the administration's objective of decentralized home-grown economic development through diverse small-scale projects that small- or mid-sized firms could handle, rather than the megaprojects long associated with downtown renewal. Hence it was considered worth the extra effort involved in soliciting bids and overseeing the engineering and roadwork, "for 50 small street resurfacing jobs, instead of several large ones that only a couple of contractors had the capacity to undertake" (Mier, 1993, p. 185). Despite the added load, under the leadership of the talented administrators who had been appointed by Mayor Washington the work proceeded apace and substantial progress was made in about a year and a half of implementation of the programs proposed in *Chicago Works Together.*

At the same time, not all of the Washington administration's programs showed such progress. With regard to linked development, there was ultimately not much to show for its efforts to levy an exaction fee on large developments to be used for funding neighborhood revitalization projects. The issue was raised forcefully in April 1984 at the first annual convention of a Chicago neighborhoods coalition known as Save Our Neighborhoods/Save Our City (SON/SOC), which brought together approximately 1,000 representatives of most of the city's White neighborhoods. At this convention a resolution was passed proposing that developers of commercial office space pay a mandatory fee of $5 per square foot over the first 100,000 square feet of new construction or substantial rehabilitation. This exaction would go to a citywide linkage fund from which funding would be distributed to each of Chicago's 77 designated community areas in proportion to population for revitalization projects to be decided by community residents. By the end of the year, Mayor Washington had met with coalition representatives and then appointed an advisory committee on linkage development. In its report to the mayor, the committee's majority proposed a somewhat different formula using a lower threshold (in excess of 50,000 rather than 100,000 square feet) and composed of a permit fee ($2 per excess square foot on receipt of building permits) plus an occupancy fee ($2 per square foot of excess occupied space over the next 4 years); nevertheless, the basic concept of exactions on over-sized development remained the same (Preston, 1989, pp. 123-126).

Unfortunately, the majority proposal to levy an exaction fee on office development found Mayor Washington caught between resistant business representatives on the advisory committee and erstwhile supporters in the Black community critical of it for their own reasons. The business members of the committee issued a minority report that opposed the majority's proposal on the basis of legal precedents that deemed exaction taxes legal "only if a 'significant and unique' relationship" could be demonstrated between the development and any local problem that it was claimed to cause. They further argued that exactions would lead to compensatory increases in rents on large commercial office developments, thereby widening the city-suburban rent gap and encouraging city businesses, particularly in the outer zone, to move to the suburbs (Preston, 1989, p. 124).

From another side, the Chicago Urban League (1986) (historically committed to Black economic and community development) critiqued the majority proposal as being vaguely drawn in regard to administrative and fiscal accountability for the neighborhood revitalization projects, as well as the criteria for project selection, and for generally being lacking in providing evidence in support of the basis for exaction. Thus the mayor found his advisory committee's report rejected both by those who supported the concept but found it weakly presented and by those who opposed it because they found it mistaken or illegal.

To extricate himself from this predicament, the mayor came up with a compromise plan in November 1985 that would have created a $10 million fund to support neighborhood economic development but that would avoid any mandatory exactions on office or other commercial development. Instead it was based partly on funds from city real estate transfer taxes and partly from developer's repayments of housing development action grant loans; finally, 70% of the total fund was to come from voluntary contributions by downtown developers (Chicago Urban League, 1986, p. 1). However, owing to a budget crisis at the time that resulted in a city lease tax highly unpopular with business groups, the mayor held off from announcing his largely voluntary contributory plan for funding neighborhood development. Amid this atmosphere of fiscal crisis, a report was issued revealing business plans for a "Super Loop" that would expand Chicago's existing downtown district (the Loop) through large-scale construction of new offices, hotels, and apartment buildings. It was estimated that if fully implemented over the better part of a decade the volume of construction involved could reverse the decline of Chicago's central district; on the other hand, a linked development tax might spoil the chances for a downtown development boom because potential investors would seek more profitable opportunities elsewhere, and Chicago's progrowth reputation would be hurt (Preston, 1989). In these circumstances, the mayor shelved his compromise plan for linked development.

This indicates that on the issue of linkage for neighborhood development, Chicago's governmental leadership, however politically innovative, was not able to bring to bear the same leverage on the business sector as it did in bringing a balky city council to pass its bond proposal for funding the repair of neighborhood infrastructure. To fully understand this, it must be kept in mind that the holders of large-scale economic capital such as corporate business interests also hold a trump card not possessed by holders of political capital such as the members of the city council. Although the political capital of the latter ultimately resides in the trust and confidence of local constituents that their representatives will in fact be responsive to their needs, the economic capital of the corporate business class is highly mobile, not tied through either custom or sentiment and surely not by prevailing conceptions of the free-market economy to any one locality. This leaves even dynamic, innovative city leaders such as Harold Washington at a considerable disadvantage. To challenge this class's relatively exclusive control of economic capital too aggressively by seeking to extract mandatory exactions from its profit-making developmental ventures is to risk being portrayed as antigrowth, even if the challenge to the holders of economic capital is on behalf of a more balanced use of that resource for the development of the community as a whole.

In short, the local holders of economic capital are, in some important respects, better able to regulate the behavior of the local holders of political capital than the reverse. At least one study of the effectiveness of linkage policies indicates that their effectiveness has been least when they are envisioned as means for basic wealth redistribution (as from downtown to general neighborhood redevelopment). Where their goals have been more limited (e.g., targeted specifically to the construction of affordable housing units) the effectiveness of linkage has been greater (Susskind, McMahon, John, & Rolley, 1986).

Yet in cases in which the locality has obtained specific commitments from developers for construction of housing for low- and moderate-income households, the results (e.g., commitments by 1985 for less than 2,700 units in San Francisco) have been "quite modest when measured against either national cuts in federal support for affordable housing or official estimates of local housing need" (Smith, 1989, pp. 95-96), especially in light of the severity of the past decade's cuts in publicly assisted housing. Thus Smith concludes that locally negotiated "linkage is clearly a poor substitute for the promotion of national policies that guarantee affordable housing as a basic need to low- and moderate-income citizens" (p. 96). This does not deny that campaigns for linkage have had considerable value in stimulating debate over local development policy and can raise public consciousness regarding the social costs of unequal public-private partnerships. Yet local linkage policies themselves cannot substitute for strong national policies targeted to unmet urban and related social needs in areas such as housing, education and job training, and economic development. However, such policies simply cannot be wished into existence. As Fainstein (1990) observed:

Without a broad national movement to support [equitable public-private development] programs . . . we must expect that local initiatives will be blocked by higher levels of government and by footloose capital that will play one locality against another. Entrepreneurship by urban progressive coalitions thus requires that they aim not only at stimulating local investment but also at building a national movement for growth with equity. (p. 44)

Clearly, such a movement would not be one-dimensionally economic in its objectives. Instead, it would direct itself intensively to realizing a new conception of urban economic development, one more inclusive of currently marginalized social group interests in restructuring American cities. In connection with this, it likely would address itself to the political and administrative issues created by New Federalism's defunding of programs for the poor and the dismantling of the organizational channels through which those programs had been delivered. But beside these concerns, such a new movement could provide a needed response to what has amounted to the removal of crucial aspects of urban policy from the public arena and from the opportunity to be informed fully about and to participate in the formation and implementation of that policy, as it affects not only the poor but also the urban working and middle classes who find themselves confronted with a changing city environment that often seems to be moving beyond their effective control.

NOTES

1. The president's report makes a point of leaving it to state governments "to encourage metropolitan-wide solutions to problems that spill over political boundaries . . . and to tackle the economic, financial, and social problems that affect the well-being of the State as it competes with others to attract and retain residents and businesses" (U.S. DHUD, 1982, p. 54). In fact, such a role may be feasible only so long as spillover problems in metropolitan areas remain confined within state boundaries. The state then appropriately would be called on as a superior level of government to resolve these problems between metropolitan counties within its own borders. But what if the jurisdictions involved straddle state lines? As Shalala and Vitullo-Martin (1989) observed, "Strong and effective as many state governments have recently become, they are often stymied in helping their cities because some 75 percent of urban areas spill across state boundaries" (p. 6). In the absence of regional institutions with authority superior to that of the states, metropolitan disputes arising from externality problems that are not readily resolvable between neighboring states are properly left for resolution by the federal government. However, it was precisely that level of government that the Reagan administration sought to withdraw from urban policy making.

2. These two tax breaks have reduced business tax liability most among the general tax incentives available since the mid-1950s, accounting for tax expenditures of over $140 billion dollars by the early 1980s (Luger, 1984, pp. 204-205).

3. For a useful introduction to Keynesian economic policy in relation to postwar American community development, see Fox (1986); for a brief conceptual discussion and related bibliography regarding the Keynesian interventionist state, see Gottdiener (1985).

4. Sharp (1990) listed several major programs used by cities (and often by states) in their competition to lure business investment: First are promotional programs, including (a) public investments in infrastructure such as bridges, streets, sewage facilities, and industrial parks; (b) developmental land management activities that acquire, consolidate, and clear parcels of land for sale, lease, or donation to developers; and (c) public relations activities such as advertising campaigns about the advantages offered by the community as a business location.

Promotional programs generally are viewed as no more than a basic minimum to provide the preconditions for economic development, often to be supplemented by financial inducement programs, which either make investment capital available at reduced cost to the investor or reduce the initial tax burden on an investment over a period of years. These programs include (a) provision of low-cost capital through issuance of federally tax-exempt bonds by states and localities on behalf of private investors for a wide variety of industrial development purposes. Such industrial development bonds [or industrial revenue bonds] effectively constitute a subsidy from the federal government to industry in the state or municipal tax jurisdiction in which they are issued; (b) provision of low-cost capital through loans, loan guarantees, and grants for development projects (including Community Development Block Grants and Urban Development Action Grants); and (c) property tax abatement, involving forgiveness on property taxes for the firm for a specified period (usually no more than 10 years).

Although numerous studies show that tax abatements and targeted business incentives generally have little net impact on business location or investment decisions (Sharp, 1990), inducement strategies have been used widely, at the least to provide symbolic evidence of a favorable business climate for investment. The growth of industrial development revenue bonds (IDRBs) as an inducement to investment has been phenomenal; during the decade of the 1970s, such bonds expanded from less than 30% of all tax-exempt securities to about 50% for new long-term issues, and small-issue IDRBs alone went "from just over $3 billion in 1978 to over $12 billion in 1981." In the case of IDRBs, the state or locality in effect transfers its tax-exempt status to the private borrower/investor, who pays off the bonded debt from the revenues of the project being funded. In allowing the exemption of these bonds from taxation, the federal treasury has "in lost tax revenue" amounting to multibillions since the early 1970s, effectively "subsidized the borrowing costs of the private investor" (Barnekov et al., 1989, pp. 76-77).

However, the original purpose of IDRBs—to stimulate economically depressed areas through the growth of industrial jobs—frequently has been skirted through the use of IDRBs by many jurisdictions for commercial projects such as "shopping centers, office buildings, and fast food outlets," often in competition with existing firms. As a consequence, private business was allowed to use federal tax expenditures as a "general investment subsidy" rather than "an instrument for regional economic development" (Barnekov et al., 1989, pp. 87, 89), with borrowed public capital flowing where it would most profit private investors, rather than to areas in greatest need of development.

5. Commenting on the "extreme complexity" that characterizes the contemporary intergovernmental system, Liner (1989) observed that "it is no simple matter to sort out state-local relationships in ordinary times, let alone to make major changes in response to the series of shock waves brought by devolution" (p. 15). Moreover, the Reagan administration added to the uncertainties implicit for cities in this increasingly complex intergovernmental system by forswearing even basic recognition of their existence, much less their different and special problems. As Farber (1989) noted, "In the view of some local officials . . . the essence—and the fundamental problem—of Reagan federalism" was that it "was so completely oriented to states that local governments had no independent identity" (p. 34).

6. State mandates that require local governments to engage in specified activities have "increasingly become an irritant in state-local relations" (Roberts, 1987, p. 55). Although various mandates may be justifiable, they have become a serious burden on local governments. Furthermore, state mandates are not the only ones that stipulate responsibilities or procedures for local governments. The federal government imposes mandates on localities as well as on states. Typically federal mandates have been requirements that those governments must satisfy as conditions for obtaining federal grants-in-aid. For local governments, the rapid growth of federal aid during the 1960s and 1970s brought with it a major expansion of mandates in areas such as civil rights, environmental pollution, occupational health and safety, and energy conservation; by 1980, there were 1,260 such federal mandates on local governments.

State mandates, on the other hand, are generally direct orders to local government (often including requirements for improvements in fiscal planning, accounting, and reporting), whether or not accompanied by state funding in assistance. They also have tended to be more numerous than federal mandates; by 1980 there were 3,415 state-local mandates (Sharp, 1990, p. 175). Thus many cities may have experienced the state mandates, which were more typically unfunded, as particularly burdensome. Moreover, with the sharp federal budgetary retrenchment of the 1980s, intergovernmental cities and other localities under fiscal stress became relatively more dependent on state aid. This led to heightened tensions between local and state governments, with cities often feeling that state governments were in various forms passing their federally mandated responsibilities for costly programs such as Medicaid as well as environmental and social programs on down to the local level.

7. Regarding the developmental partnerships characteristic of the 1970s and 1980s, Levine (1989) observed that these have been generally "insulated from public influence and dominated by downtown-oriented interests." Often taking the form of quasi-public development corporations, these partnerships have emphasized "deal-making and profit opportunities," rather than systematic planning of how best to deploy public resources to create good jobs and meet pressing neighborhood needs." Typically, the insulation of the development corporations has been used "to keep those groups deriving few benefits from partnership activity from interfering with the deal-making process" (p. 27). For debate over the degree to which developmental programs have addressed the needs of neighborhoods as well as those of the downtown business

district in a restructuring city often cited as a model, see Levine (1987a, 1987b) and Berkowitz (1987). For relevant discussion regarding the political-economic context and implications of urban developmental policy, see Sanders and Stone (1987a, 1987b) and Peterson (1987).

8. *Community development corporation* is a generic term for all nonprofit community-based organizations involved in development activity, typically in low-income communities. Originated in the late 1960s during the War on Poverty, economic development activities of CDCs nationally (based on a survey of 834 CDCs) in just the 5-year period between 1984 and 1989 included housing rehabilitation and construction (125,000 dwelling units), business loans (2,048), equity investments in business ventures (218), ownership of 427 businesses, and creation of almost 90,000 jobs (National Congress for Community Economic Development, 1989, p. 1). However, these economic accomplishments can be evaluated fully only in relation to several less readily measurable, but no less important social and organizational aspects of community economic development, such as community participation in the development process, the development of leadership capacities, and the general impact on community expectations and capacities for further development activities. These less tangible elements of the economic development process in fact have been identified consistently with the most effective CDCs—those with most capability to sustain and complete development projects (Betancur, Bennett, & Wright, 1991).

In this perspective, truly effective CDCs are those that develop the social-organizational and political capacities for community development, rather than simply focusing on economic projects as quick fixes. Experienced analysts of the Committee for Economic Development observe that only CDCs that combine these social-organizational capacities with economic development activities can attain true empowerment over the development process in their communities. But under the pressures of federal retrenchment, many CDCs have moved away from this more comprehensive approach to an emphasis on measurable economic payoff alone. Clearly, this tendency to emphasize quantitative economic results needs to be corrected if "CDCs are to become more than marginal players" in shaping community development (Betancur, Bennett, & Wright, 1991, pp. 200, 201). Thus "the movement which spawned CDCs must be reformed and reinvigorated" (Keating & Krumholz, 1988, p. 16), so that "physical development and people development" (as well as economic and political development) all can occur "each at its own pace, each with its own integrity" (Shiffman & Motley, 1989, p. 36).

At the local level, Chicago—as much as any older industrial city—exemplifies the preconditions for such a movement. Against the background of a long history of labor and community activism, Chicago community-based organizations (CBOs) rallied after the recession of the mid-1970s and under the Reaganist New Federalism of the 1980s to join forces in coalitions such as the Chicago Association of Neighborhood Development Organizations (CANDO), established in 1978 to provide information and technical assistance on neighborhood development; to support community-based organizing, planning, and advocacy; and to lobby government for economic development programs (Mier, Wiewel, & Alpern, 1993).

By the late 1980s CANDO, together with the Chicago Workshop on Economic Development (CWED, established in 1982), and several other community-based coalitions (the Chicago Jobs Council, the Chicago Rehab Network, the Community Land Use Network, and the Neighborhood Capital Budget Group) had joined in a coalition of coalitions known as the "Neighborhood Agenda." The objective was to engage in bottom-up agenda setting to influence city budgeting on behalf of major capital investment in a broad spectrum of low- and moderate-income White-ethnic and minority neighborhoods. These coalitions, reflecting "the increased strength and sophistication of the neighborhood development movement" in Chicago by the early 1990s, appear to have played a significant role in persuading Mayor Richard M. Daley (son of the famous Mayor Daley who symbolized machine government in post-World War II

Chicago) not to abandon Harold Washington's policy of investment in neighborhoods or his support for retention of industrial manufacturing in planned manufacturing districts. The visible result was to be seen in the commitment of Daley and the city council to "a $160-million infrastructure bond issue, a start toward the Agenda's $400-million program." To this point, the less tangible result seems to have been the survival of decentralized, neighborhood-oriented policy making in the post-Washington era, of which Mier et al. (1993) observed that "a process that has now been institutionalized for almost 10 years will likely have formed strong enough roots that it can no longer be extirpated easily" (pp. 131, 133).

References

Advisory Commission on Intergovernmental Relations (ACIR). (1976). *Improving urban America: A challenge to federalism.* Washington, DC: Government Printing Office.

Advisory Commission on Intergovernmental Relations (ACIR). (1977). *Improving federal grants management.* Washington, DC: Government Printing Office.

Advisory Commission on Intergovernmental Relations (ACIR). (1982). *The federal role in the federal system: The dynamics of growth: Reducing unemployment: Intergovernmental dimensions of a national problem.* Washington, DC: Government Printing Office.

Alcaly, R. E., & Mermelstein, D. (Eds.). (1976). *The fiscal crisis of American cities.* New York: Vintage.

Allensworth, D. T. (1975). *The political realities of urban planning.* New York: Praeger.

Altshuler, A. (1970). *Community control: The black demand for participation in large American cities.* New York: Pegasus.

Anderson, M. (1964). *The federal bulldozer.* Cambridge: MIT Press.

Anton, T. (Ed.). (1984). Intergovernmental change in the United States: An assessment of the literature. In T. Miller (Ed.), *Public sector performance* (pp. 15-64). Baltimore, MD: Johns Hopkins University Press.

Ashton, P. J. (1984). Urbanization and the dynamics of suburban development under capitalism. In W. K. Tabb & L. Sawers (Eds.), *Marxism and the metropolis: New perspectives in urban political economy* (2nd ed., pp. 54-81). New York: Oxford University Press.

Banfield, E. C. (1970). *The unheavenly city.* Boston: Little, Brown.

Banfield, E. C., & Wilson, J. Q. (1963). *City politics.* New York: Vintage.

Baran, P. A., & Sweezey, P. M. (1966). *Monopoly capital.* New York: Monthly Review Press.

Barnekov, T., Boyle, R., & Rich, D. (1989). *Privatism and urban policy in Britain and the United States.* Oxford, UK: Oxford University Press.

Beer, S. (1976). The adoption of general revenue sharing: A case study in public sector politics. *Public Policy, 24,* 127-195.

Bell, E., & Held, V. (1969). The community revolution. *The Public Interest, 16,* 142-177.

Bellush, J., & Hausknecht, M. (Eds.). (1967). *Urban renewal: People, politics and planning.* Garden City, NY: Doubleday.

Bendick, M., & Rasmussen, D. W. (1986). Enterprise zones and inner-city economic revitalization. In G. E. Peterson & C. W. Lewis (Eds.), *Reagan and the cities* (pp. 97-129). Washington, DC: Urban Institute.

Benson, J. K. (1975). The interorganizational network as a political economy. *Administrative Science Quarterly, 20,* 229-249.

Benson, J. K. (1982). A framework for policy analysis. In D. I. Rogers & D. A. Whetten (Eds.), *Interorganizational coordination* (pp. 137-176). Ames: Iowa State University Press.

Berkowitz, B. L. (1987). Rejoinder to downtown redevelopment as an urban growth strategy: A critical appraisal of the Baltimore renaissance. *Journal of Urban Affairs, 9,* 125-132.

Berman, P. (1978). The study of macro and micro-implementation. *Public Policy, 26,* 157-184.

Berry, B.J.L., & Kasarda, J. D. (1977). *Contemporary urban ecology.* New York: Macmillan.

Betancur, J. J., Bennett, D. E., & Wright, P. A. (1991). Effective strategies for community economic development. In P. W. Nyden & W. Wiewel (Eds.), *Challenging uneven development: An urban agenda for the 1990s* (pp. 198-224). New Brunswick, NJ: Rutgers University Press.

Boesel, A. W. (1974). Local personnel management: Organizational problems and operating practices. In International City Management Association (Ed.), *Municipal year book* (pp. 82-94). Washington, DC: International City Management Association.

Bollens, J. G., & Schmandt, H. J. (1970). *The metropolis: Its people, politics, and economic life* (2nd ed.). New York: Harper & Row.

Bowie, L. (1992, May 10). Life-science push faces big hurdles. *Baltimore Sun,* pp. D1, D4.

Bradford, C. P., & Rubinowitz, L. F. (1975). The urban-suburban disinvestment process. *American Academy of Political and Social Science Annals, 422,* 77-86.

Breckenfeld, G. (1977). How cities can cope with shrinkage. In U.S. House of Representatives, *How cities can grow old gracefully: Documents for the house committee on banking, finance, and urban affairs* (pp. 105-117). 95th Congress, 1st session. Washington, DC: Government Printing Office.

Brown, L. D., Fossett, J. W., & Palmer, K. T. (1984). *The changing politics of federal grants.* Washington, DC: Brookings Institution.

Burchell, R. W., Carr, J. H., Florida, R. L., Nemeth, J., Pawlik, M., & Barreto, F. R. (1984). *The new reality of municipal finance: The rise and fall of the intergovernmental city.* New Brunswick, NJ: Center for Urban Policy Research.

Burgess, E. W. (1967). The growth of the city. In R. E. Park & E. W. Burgess (Eds.), *The city* (pp. 47-62). Chicago: University of Chicago Press. (Original work published 1925)

Candeub, I. (1970). New techniques in making the general plan. In E. Erber (Ed.), *Urban planning in transition* (pp. 216-224). New York: Grossman.

Carnoy, M., Shearer, D., & Rumberger, R. (1983). *A new social contract.* New York: Harper & Row.

Chandler, L. V. (1970). *America's greatest depression, 1929-1941.* New York: Harper & Row.

Chicago Urban League. (1986). *The policy issue linked development: A preliminary report to the board of directors.* Chicago: Author.

Chicago Works Together Planning Task Force. (1984). *Chicago works together.* Chicago: Author.

Civilian Production Administration, Industrial Statistics Division. (1946). *War-time manufacturing plant expansion.* Washington, DC: Author.

Clark, T. B., Iglehart, J. K., & Lilley, W., III. (1972a, October 7). Federalism report: Revenue sharing bill authorizes sweeping innovations in federal aid system. *National Journal,* pp. 1553-1566.

Clark, T. B., Iglehart, J. K., & Lilley, W., III. (1972b, December 16). New federalism I: Return of power to states and cities looms as theme of Nixon's second-term domestic policy. *National Journal,* pp. 1908-1912.

Cloward, R. A., & Ohlin, L. E. (1960). *Delinquency and opportunity.* Glencoe, IL: Free Press.

Coalition for Low-Income Community Development (CL-ICD). (1993). *CDBG stories: An organizing manual.* Baltimore, MD: Author.

Conlan, T. (1988). *New federalism: Intergovernmental reform from Nixon to Reagan.* Washington, DC: Brookings Institution.

Constable, P. (1980, June 15). "Passport" from ghetto assailed. *Baltimore Sun,* pp. B1, B2.

Cook, K. S. (1977). Exchange and power in networks in interorganizational relations. *Sociological Quarterly, 18,* 62-82.

Cuciti, P. L. (1990). A nonurban policy: Recent policy shifts affecting cities. In M. Kaplan & F. James (Eds.), *The future of national urban policy* (pp. 235-250). Durham, NC: Duke University Press.

Cummings, C., Koebel, T., Whitt, J. A. (1989). Redevelopment in downtown Louisville. In G. D. Squires (Ed.), *Unequal partnerships* (pp. 202-221). New Brunswick, NJ: Rutgers University Press.

DeGrove, J. M., & Brumback, B. C. (1985). State-local partnerships. In C. R. Warren (Ed.), *Urban policy in a changing federal system* (pp. 202-225). Washington, DC: National Academy Press.

Deleon, R., & LeGates, R. (1976). *Redistribution effects of special revenue sharing for community development.* Berkeley: University of California, Institute of Governmental Studies.

Dolbeare, K. M., & Edelman, M. J. (1985). *American politics: Policies, power, and change.* Lexington, MA: D. C. Heath.

Dommel, P. R., Hall, J. S., Bach, V. E., Rubinowitz, L., Haley, L. L., & Jackson, J. S. (1982). *Decentralizing urban policy.* Washington, DC: Brookings Institution.

Duncan, O. D. (1959). Human ecology and population studies. In P. M. Hauser & O. D. Duncan (Eds.), *The study of populations* (pp. 678-716). Chicago: University of Chicago Press.

Duncan, O. D. (1964). Social organization and the ecosystem. In R.E.L. Faris (Ed.), *Handbook of modern sociology* (pp. 37-82). Chicago: Rand McNally.

Ehrenreich, B., & Ehrenreich, J. (1979). The professional-managerial class. In P. Walker (Ed.), *Between labor and capital* (pp. 5-45). Boston: South End Press.

Esping-Andersen, G., Friedland, R., & Wright, E. O. (1976). Modes of class struggle and the capitalist state. *Kapitalistate: Working Papers on the Capitalist State, 4-5,* 186-220.

Fainstein, S. (1990). The changing world economy and urban restructuring. In D. Judd & M. Parkinson (Ed.), *Leadership and urban regeneration: Cities in North America and Europe* (pp. 31-47). Newbury Park, CA: Sage.

Fainstein, S. S., Fainstein, N. I., Hill, R. C., Judd, D., Smith, M. P., Armistead, P. J., & Keller, M. (1983). *Restructuring the city: The political economy of urban development.* New York: Longman.

Farber, S. B. (1989). Federalism and state-local relations. In E. B. Liner (Ed.), *A decade of devolution* (pp. 27-50). Washington, DC: Urban Institute.

Federal Register. (1977, October 25). Vol. 42. Washington, DC: Office of the Federal Register.

Fossett, J. W. (1984). The politics of dependence: Federal aid to big cities. In L. D. Brown, J. W. Fossett, & K. T. Palmer (Eds.), *The changing politics of federal grants* (pp. 108-165). Washington, DC: Brookings Institution.

Fox, D. M. (Ed.). (1972). *The new urban politics: Cities and the federal government.* Pacific Palisades, CA: Goodyear.

Fox, K. (1986). *Metropolitan America: Urban life and urban policy in the United States, 1940-1980.* Jackson: University of Mississippi Press.

Fried, J. (1971). *Housing crisis U.S.A.* New York: Praeger.

Frieden, B., & Kaplan, M. (1975). *The politics of neglect.* Cambridge: MIT Press.

Frieden, B., & Kaplan, M. (1976). *Community development and the model cities legacy* (Working Paper No. 42). Cambridge, MA: Joint Center for Urban Studies.

Frieden, B., & Sagalyn, L. (1989). *Downtown, inc.: How America rebuilds cities.* Cambridge: MIT Press.

Friedland, R. (1982). *Power and crisis in the city.* London: Macmillan.

Friedland, R., Piven, F. F., & Alford, R. R. (1984). Political conflict, urban structure and the fiscal crisis. In W. K. Tabb & L. Sawers (Eds.), *Marxism and the metropolis: New perspectives in urban political economy* (pp. 273-298). New York: Oxford University Press.

Friedman, L. M. (1967). Government and slum housing: Some general considerations. *Law and Contemporary Problems, 32*, 357-370.

Friedman, L. M. (1977). The social and political context of the war on poverty: An overview. In R. H. Haveman (Ed.), *A decade of federal antipoverty programs* (pp. 21-47). New York: Academic Press.

Galbraith, J. K. (1958). *The affluent society.* Boston: Houghton Mifflin.

Gans, H. (1970). The need for planners trained in policy formulation. In E. Erber (Ed.), *Urban planning in transition* (pp. 239-246). New York: Grossman.

Gantz, A. (1972). *Our large cities: New light on their recent transformation.* Cambridge: MIT Laboratory for Environmental Studies.

Gelfand, M. I. (1975). *A nation of cities: The federal government and urban America, 1933-1965.* New York: Oxford University Press.

Gilbert, N. (1970). *Clients or constituents.* San Francisco: Jossey-Bass.

Gilbert, N. (1974). *Dimensions of social welfare planning.* Englewood Cliffs, NJ: Prentice Hall.

Gilbert, N., & Specht, H. (1977). *Dynamics of community planning.* Cambridge, MA: Ballinger.

Glaab, C. N., & Brown, A. T. (1967). *A history of urban America.* New York: Macmillan.

Glickman, N. J. (1984). Economic policy and the cities: In search of Reagan's "real" urban policy. *Journal of the American Planning Association, 50*, 471-478.

Gold, D. (1975). James O'Connor's "The fiscal crisis of the state": An overview. In D. Mermelstein (Ed.), *The economic crisis reader* (pp. 123-131). New York: Vintage.

Goldfield, D. R., & Brownell, B. A. (1979). *Urban America: From downtown to no town.* Boston: Houghton Mifflin.

Gordon, D. M. (1984). Capitalist development and the history of American cities. In W. K. Tabb & L. Sawers (Eds.), *Marxism and the metropolis: New perspectives in urban political economy* (pp. 21-53). New York: Oxford University Press.

Gottdiener, M. (1985). *The social production of urban space.* Austin: University of Texas Press.

Gramlich, E. (1994). Missing the target: A brief history of CDBG's failure to focus on low-income people. *Targeting Times, 5,* 1, 13-15.

Greer, S. (1962). *The emerging city.* New York: Free Press.

Hahn, H., & Levine, C. H. (Eds.). (1984). *Readings in urban politics* (2nd ed.). New York: Longman.

Haider, D. H. (1974). *When governments come to Washington.* New York: Free Press.

Hanson, R. (Ed.). (1983). *Rethinking urban policy: Urban development in an advanced economy.* Washington, DC: National Academy Press.

Harrigan, J. J. (1985). *Political change in the metropolis* (3rd ed.). Boston: Little, Brown.

Harrington, M. (1962). *The other America.* New York: Macmillan.

Hartman, C. W. (1975). *Housing and social policy.* Englewood Cliffs, NJ: Prentice Hall.

Harvey, D. (1978). Labor, capital and class struggle around the built-environment in advanced capitalist societies. In K. R. Cox (Ed.), *Urbanization and conflict in market societies* (pp. 9-35). Chicago: Maaroufa.

Hawley, A. (1950). *Human ecology: A theory of community structure.* New York: Ronald Press.

Hawley, A. (1971). *Urban society: An ecological approach.* New York: Ronald Press.

Hawley, W. D., & Rogers, D. (Eds.). (1974). *Improving the quality of urban management.* Beverly Hills, CA: Sage.

Hays, R. A. (1985). *The federal government and urban housing.* New York: State University of New York Press.

Hays, S. P. (1984). The politics of reform in municipal government in the progressive era. In H. Hahn & C. H. Levine (Eds.), *Readings in urban politics* (pp. 54-71). New York: Longman.

Henig, J. R. (1985). *Public policy and federalism: Issues in state and local politics.* New York: St. Martin's Press.

Hill, R. C. (1984). Fiscal crisis, austerity politics, and alternative urban policies. In W. K. Tabb & L. Sawers (Eds.), *Marxism and the metropolis: New perspectives in urban political economy* (pp. 298-322). New York: Oxford University Press.

Hofstadter, R. (1955). *The age of reform.* New York: Knopf.

Holleb, D. B. (1975). The direction of urban change. In H. S. Perloff (Ed.), *Agenda for the new urban era* (pp. 11-43). Chicago: American Society of Planning Officials.

Hollingshead, A. (1961). A re-examination of ecological theory. In G. A. Theodorson (Ed.), *Studies in human ecology* (pp. 108-114). New York: Harper & Row.

Howitt, A. M. (1984). *Managing federalism: Studies in intergovernmental relations.* Washington, DC: Congressional Quarterly Press.

Humphrey, H. H. (1966). The war on poverty. In R. C. Everett (Ed.), *Anti-poverty programs* (pp. 6-17). Dobbs Ferry, NY: Oceana.

Huthmacher, J. J. (1968). *Senator Wagner and the rise of urban liberalism.* New York: Atheneum.

Ilchman, W. F., & Uphoff, N. T. (1969). *The political economy of change.* Berkeley: University of California Press.

Jacobs, B. G., Harney, K. R., Edson, C. L., & Lane, B. S. (1982). *Guide to federal housing programs.* Washington, DC: The Bureau of National Affairs, Inc.

Jacobs, S. S., & Roistacher, E. A. (1980). The urban impacts of HUD's urban development action grant program. In N. J. Glickman (Ed.), *The urban impacts of federal policies* (pp. 335-362). Baltimore, MD: Johns Hopkins University Press.

James, J. (1972). Federalism and the model cities program. *Publius, 2,* 69-94.

Jansson, B. S. (1988). *The reluctant welfare state.* Belmont, CA: Wadsworth.

Joint Economic Committee. (1982, January 14). *Congress, emergency interim survey: Fiscal conditions of 48 large cities.* Washington, DC: Government Printing Office.

Jones, E. (1966). *Towns and cities.* London: Oxford University Press.

Judd, D. R. (1988). *The politics of American cities* (3rd ed.). Glenview, IL: Scott, Foresman.

Judd, D. R., & Ready, R. L. (1986). Entrepreneurial cities and the new politics of economic development. In G. E. Peterson & C. W. Lewis (Eds.), *Reagan and the cities* (pp. 209-247). Washington, DC: Urban Institute.

Kain, J. F. (1970). The distribution and movement of jobs and industry. In J. Q. Wilson (Ed.), *The metropolitan enigma* (pp. 1-43). Garden City, NY: Doubleday.

Kantor, P. (1988). *The dependent city: The changing political economy of urban America.* Glenview, IL: Scott, Foresman.

Kantor, P. (1991). A case for a national urban policy. *Urban Affairs Quarterly, 26,* 394-415.

Kasarda, J. D. (1978). Urbanization, community, and the metropolitan problem. In D. S. Street (Ed.), *Handbook of contemporary urban life* (pp. 27-57). San Francisco: Jossey-Bass.

Kasarda, J. D. (1980). Implications of contemporary redistribution trends for national urban policy. *Social Science Quarterly, 61,* 373-400.

Kasarda, J. D. (1985). Urban change and minority opportunities. In P. E. Peterson (Ed.), *The new urban reality* (pp. 33-67). Washington, DC: Brookings Institution.

Katz, M. B. (1989). *The undeserving poor.* New York: Pantheon.

Katznelson, I. (1976). The crisis of the capitalist city: Urban politics and social control. In W. D. Hawley & M. Lipsky (Eds.), *Theoretical perspectives on urban politics* (pp. 214-229). Englewood Cliffs, NJ: Prentice Hall.

Katznelson, I. (1981). *City trenches.* Chicago: University of Chicago Press.

Keating, W. D. (1986). Linking downtown development to broader community goals. *Journal of the American Planning Association, 52,* 133-141.

Keating, W. D., & Krumholz, N. (1988). *Community development corporations in the United States: Their role in housing and urban development.* Cleveland, OH: Cleveland State University.

Keller, R. (1991, May 26). Directing Baltimore's economic development efforts toward life sciences. *Baltimore Sun,* p. N5.

Kettl, D. F. (1979). Can the cities be trusted? *Political Science Quarterly, 94,* 437-451.

Kettl, D. F. (1980). *Managing community development in the new federalism.* New York: Praeger.

Kettl, D. F. (1988). *Government by proxy: (Mis?)managing federal programs.* Washington, DC: Congressional Quarterly Press.

Kirkland, E. C. (1970). Urban growth and industrial development. In A. M. Wakstein (Ed.), *The urbanization of America* (pp. 212-222). Boston: Houghton Mifflin.

Kirlin, J. J. (1978). Adapting the intergovernmental fiscal system to the demands of an advanced economy. In G. Tobin (Ed.), *The changing structure of the city* (pp. 77-103). Beverly Hills, CA: Sage.

Knoke, D. (1990). *Political networks: The structural perspective.* Cambridge, UK: Cambridge University Press.

Kotler, M. (1969). *Neighborhood government: The local foundations of political life.* Indianapolis: Bobbs-Merrill.

Kravitz, S. (1969). The community action program—past, present and its future? In J. L. Sundquist (Ed.), *On fighting poverty* (pp. 52-70). New York: Basic Books.

Lachman, M. L. (1975). Planning for community development: A proposed approach. *Journal of Housing, 32,* 58-62.

Larson, C. J., & Nikkel, S. R. (1979). *Urban problems: Perspectives on corporations, governments, and cities.* Boston: Allyn & Bacon.

Laumann, E. O., Galaskiewicz, J., & Mardsen, P. U. (1978). Community structure as interorganizational linkages. *Annual Review of Sociology, 4,* 455-484.

Laumann, E. O., & Knoke, D. (1987). *The organizational state: Social choice in national policy domains.* Madison: University of Wisconsin Press.

Levi, M. (1974). Poor people against the state. *Review of Radical Political Economics, 6,* 76-98.

Levine, M. A. (1983). The Reagan urban policy. *Journal of Urban Affairs, 5,* 17-28.

Levine, M. V. (1987a). Downtown redevelopment as an urban growth strategy: A critical appraisal of the Baltimore renaissance. *Journal of Urban Affairs, 9,* 103-123.

Levine, M. V. (1987b). Response to Berkowitz, economic development in Baltimore: Some additional perspectives. *Journal of Urban Affairs, 9,*133-138.

Levine, M. V. (1989). The politics of partnership: Urban redevelopment since 1945. In G. D. Squires (Ed.), *Unequal partnerships* (pp. 12-34). New Brunswick, NJ: Rutgers University Press.

Levine, S., & White, P. (1961). Exchange as a conceptual framework for the study of interorganizational relationships. *Administrative Science Quarterly, 5,* 583-601.

Levitan, S. (1969). *The great society's poor law: A new approach to poverty.* Baltimore, MD: Johns Hopkins University Press.

Levitan, S. A., & Taggart, R. C. (1976). *The promise of greatness.* Cambridge, MA: Harvard University Press.

Lewis, E. (1973). *The urban political system.* Hinsdale, IL: Dryden.

Lindbloom, C. G., & Farrah, M. (1968). *The citizen's guide to urban renewal.* West Trenton, NJ: Chandler-Davis.

Liner, E. B. (1989). Sorting out state-local relations. In E. B. Liner (Ed.), *A decade of devolution* (pp. 3-25). Washington, DC: Urban Institute.

Long, N. E. (1971). The city as reservation. *The Public Interest, 25,* 22-38.

Lowi, T. J. (1967). Machine politics: Old and new. *The Public Interest, 9,* 83-92.

Luce, T., & Pack, J. R. (1984). State support under the new federalism. *Journal of Policy Analysis and Management, 3,* 339-358.

Luger, M. I. (1984). Federal tax incentives as industrial and urban policy. In L. Sawers & W. K. Tabb (Eds.), *Sunbelt/snowbelt* (pp. 201-234). New York: Oxford University Press.

Magill, R. S. (1979). *Community decision-making for social welfare.* New York: Human Sciences Press.

Mandel, E. (1970). *Marxist economic theory.* New York: Monthly Review Press.

Markusen, A. R. (1980). Urban impact analysis: A critical forecast. In N. J. Glickman (Ed.), *The urban impact of federal policies* (pp. 103-118). Baltimore, MD: Johns Hopkins University Press.

Martin, R. C. (1965). *The cities and the federal system.* New York: Atherton.

Marx, K., & Engels, F. (1967). *Communist manifesto.* Moscow: Progress.

McKean, E. C., & Taylor, H. C. (1955). *Public works and employment.* Chicago: Public Administration Service.

Mier, R. (1993). Community development and diversity. In R. Mier (Ed.), *Social justice and local development policy* (pp. 182-199). Newbury Park, CA: Sage.

Mier, R., Moe, K. J., & Sherr, I. (1986). Strategic planning and the pursuit of reform, economic development and equity. *Journal of the American Planning Association, 52,* 299-309.

Mier, R., Wiewel, W., & Alpern, L. (1993). Decentralization of policy making under Mayor Harold Washington. In R. Mier (Ed.), *Social justice and local development policy* (pp. 115-134). Newbury Park, CA: Sage.

Milward, H. B. (1982). Interorganizational policy systems and research on public organizations. *Administration & Society, 13,* 456-478.

Miringoff, M. L., & Opdycke, S. (1986). *American social welfare policy: Reassessment and reform.* Englewood Cliffs, NJ: Prentice Hall.

Mollenkopf, J. (1983). *The contested city.* Princeton, NJ: Princeton University Press.

Molotch, H. (1976). The city as a growth machine. *American Journal of Sociology, 82,* 309-330.

Morris, M. (1976). *New federalism and community development.* Washington, DC: Joint Center for Political Studies.

Moynihan, D. P. (1969). *Maximum feasible misunderstanding.* New York: Free Press.

Muller, P. O. (1981). *Contemporary suburban America.* Englewood Cliffs, NJ: Prentice Hall.

Mumford, L. (1961). *The city in history.* New York: Harcourt, Brace, & World.

Nader Congress Project. (1975). *The money committees.* New York: Grossman.

Nathan, R. P., & Adams, C. (1977). *Revenue sharing: The second round.* Washington, DC: Brookings Institution.

Nathan, R. P., Dommel, P. P., Liebschutz, S. J., Morris, M. D., & Associates. (1977). *Block grants for community development.* Washington, DC: U.S. Department of Housing and Urban Development.

Nathan, R. P., & Doolittle, F. C. (Eds.). (1987). *Reagan and the states.* Princeton, NJ: Princeton University Press.

National Commission on Urban Problems. (1968). *Building the American city.* Washington, DC: Government Printing Office.

National Congress for Community Economic Development. (1989). *Against all odds: The achievements of community-based development organizations.* Washington, DC: Author.

O'Connor, J. (1973). *The fiscal crisis of the state.* New York: St. Martin's.

O'Donnell, P. (1977). Industrial capitalism and the rise of modern American cities. *Kapitalistate: Working Papers on the Capitalist State, 6,* 91-128.

Pagano, M. A., & Moore, R.J.T. (1985). *Cities and fiscal choices.* Durham, NC: Duke University Press.

Palmer, K. T. (1984). The evolution of grant policies. In L. D. Brown, J. W. Fossett, & K. T. Palmer (Eds.), *The changing politics of federal grants* (pp. 5-54). Washington, DC: Brookings Institution.

Park, R. E. (1952). *Human communities.* New York: Free Press.

Park, R. E., & Burgess, E. W. (Eds.). (1925). *The city.* Chicago: University of Chicago Press.

Parkinson, M., Foley, B., & Judd, D. R. (Eds.). (1989). *Regenerating the cities.* Glenview, IL: Scott, Foresman.

Perlman, J. (1979). Grassrooting the system. In F. M. Cox, J. L. Erlich, J. Rothman, & J. E. Tropman (Eds.), *Strategies of community organizations* (pp. 403-425). Itasca, IL: Peacock.

Peterson, P. (1981). *City limits.* Chicago: University of Chicago Press.

Peterson, P. E. (1987). Analyzing developmental politics: A response to Sanders and Stone. *Urban Affairs Quarterly, 22,* 540-547.

Peterson, P. E., & Greenstone, J. D. (1977). The mobilization of low-income communities through community action. In R. H. Haveman (Ed.), *A decade of federal antipoverty programs* (pp. 241-279). New York: Academic Press.

Peterson, P. E., Rabe, B. G., & Wong, K. K. (1986). *When federalism works.* Washington, DC: Brookings Institution.

Piven, F. F. (1974a). The great society as political strategy. In R. A. Cloward & F. F. Piven (Eds.), *The politics of turmoil* (pp. 271-283). New York: Vintage.

Piven, F. F. (1974b). The new urban programs: The strategy of federal intervention. In R. A. Cloward & F. F. Piven (Eds.), *The politics of turmoil* (pp. 284-313). New York: Vintage.

Piven, F. F. (1976). The urban crisis: Who got what and why. In R. E. Alcaly & D. Mermelstein (Eds.), *The fiscal crisis of American cities* (pp. 132-144). New York: Vintage.

Piven, F. F., & Cloward, R. A. (1977). *Poor people's movements.* New York: Pantheon.

Piven, F. F., & Cloward, R. A. (1982). *The new class war.* New York: Pantheon.

Plotnick, R. D., & Skidmore, F. (1975). *Progress against poverty.* New York: Academic Press.

Poulantzas, N. (1972). The problem of the capitalist state. In R. Blackburn (Ed.), *Ideology in social science* (pp. 238-262). London: Fontana.

Pratter, J. S. (1977). Strategies for city investment. In U.S. House of Representatives, *How cities can grow old gracefully: Documents for the house committee on banking, finance, and urban affairs* (pp. 79-90). 95th Congress, 1st session. Washington, DC: Government Printing Office.

Pred, A. (1977). *City systems in advanced economies.* New York: John Wiley.

Pressman, J. L. (1985). Federal programs and city politics. In L. J. O'Toole (Ed.), *American intergovernmental relations* (pp. 119-123). Washington, DC: Congressional Quarterly Press.

Pressman, J. L., & Wildavsky, A. B. (1979). *Implementation* (2nd ed.). Berkeley: University of California Press.

Preston, M. B. (1989). The politics of economic redistribution in Chicago. In M. Parkinson, B. Foley, & D. R. Judd (Eds.), *Regenerating the cities* (pp. 117-128). Glenview, IL: Scott, Foresman.

Reagan, M., & Sanzone, J. (1981). *The new federalism* (2nd ed.). New York: Oxford University Press.

Rees, P. M. (1972). Problems of classifying sub-areas within cities. In B. J. L. Berry (Ed.), *City classification hand book* (pp. 265-330). New York: Wiley-Interscience.

Ripley, R. B. (1972). *The politics of economic and human resource development.* Indianapolis: Bobbs-Merrill.

Roberts, J. (1987). State actions affecting local governments. In International City Management Association (Ed.), *Municipal year book* (pp. 54-64). Washington, DC: International City Management Association.

Robertson, D. B., & Judd, D. R. (1989). *The development of American public policy: The structure of policy restraint.* Glenview, IL: Scott, Foresman.

Rondinelli, D. A. (1975). *Urban and regional developmental planning: Policy and administration.* Ithaca, NY: Cornell University Press.

Rose, S. (1977). The transformation of community action. In R. L. Warren (Ed.), *New perspectives in the American community* (pp. 471-481). Chicago: Rand McNally.

Rosenfeld, R. A. (1980). Who benefits and who decides? The uses of community development block grants. In D. B. Rosenthal (Ed.), *Urban revitalization* (pp. 211-235). Beverly Hills, CA: Sage.

Salisbury, R. (1964). Urban politics: The new convergence of power. *Journal of Politics, 26,* 775-797.

Sanders, H. T., & Stone, C. N. (1987a). Developmental politics reconsidered. *Urban Affairs Quarterly, 22,* 521-539.

Sanders, H. T., & Stone, C. N. (1987b). Competing paradigms: A rejoinder to Peterson. *Urban Affairs Quarterly, 22,* 548-551.

Savas, E. S. (1983). A positive urban policy for the future. *Urban Affairs Quarterly, 18,* 447-453.

Savitch, H. V. (1979). *Urban policy and the exterior city.* New York: Pergamon.

Schultze, C. L. (1977). *The public use of private interest.* Washington, DC: Brookings Institution.

Scott, J. (1969). Corruption, machine politics, and political change. *American Political Science Review, 63,* 1142-1158.

Scott, W. R., & Meyer, J. W. (1991). The organization of societal sectors: Propositions and early evidence. In W. W. Powell & P. J. Dimaggio (Eds.), *The new institutionalism in organizational analysis* (pp. 108-140). Chicago: University of Chicago Press.

Shalala, D. E., & Vitullo-Martin, T. (1989). Rethinking the urban crisis. *Journal of the American Planning Association, 55,* 3-13.

Shank, A., & Conant, R. W. (1975). *Urban perspectives: Politics and policies.* Boston: Holbrook.

Shannon, T. (1983). *Urban problems in sociological perspective.* Prospect Heights, IL: Waveland.

Sharp, E. B. (1990). *Urban politics and administration.* New York: Longman.

Shearer, D. (1989). In search of equal partnerships. In G. D. Squires (Ed.), *Unequal partnerships* (pp. 289-307). New Brunswick, NJ: Rutgers University Press.

Shefter, M. (1976). The emergence of the political machine: An alternative view. In W. D. Hawley & M. Lipsky (Eds.), *Theoretical perspectives on urban politics* (pp. 14-44). Englewood Cliffs, NJ: Prentice Hall.

Shevky, E., & Bell, W. (1955). *Social area analysis.* Stanford, CA: Stanford University Press.

Shiffman, R., & Motley, S. (1989). *Comprehensive and integrative planning for community development.* New York: Pratt Institute.

Smith, M. P. (1988). *City, state, and market.* New York: Blackwell.

Smith, M. P. (1989). The uses of linked-development policies in U.S. cities. In M. Parkinson, B. Foley, & D. R. Judd (Eds.), *Regenerating the cities* (pp. 85-99). Glenview, IL: Scott, Foresman.

Smith, M. P., & Judd, D. R. (1984). American cities: The production of ideology. In M. P. Smith (Ed.), *Cities in transformation* (pp. 173-196). Beverly Hills, CA: Sage.

Solomon, A. P. (1980). The emerging metropolis. In A. P. Solomon (Ed.), *The prospective city* (pp. 3-28). Cambridge: MIT Press.

Squires, G. D. (Ed.). (1989). *Unequal partnerships.* New Brunswick, NJ: Rutgers University Press.

Stanfield, R. L. (1981, March 21). Economic development aid: Shell game or key to urban rejuvenation? *National Journal,* pp. 494-497.

Starr, R. (1978). Making New York smaller. In G. Sternlieb & J. W. Hughes (Eds.), *Revitalizing the northeast* (pp. 378-389). New Brunswick, NJ: Rutgers University Press.

Sternlieb, G. L. (1971). The city as sandbox. *The Public Interest, 25,* 14-21.

Sternlieb, G., & Hughes, J. W. (Eds.). (1975). *Post-Industrial America: Metropolitan decline and inter-regional job shifts.* New Brunswick, NJ: Center for Urban Policy Research.

Stone, C. N., & Sanders, H. T. (1987). *Politics of Urban Development.* Lawrence: University Press of Kansas.

Sundquist, J. L. (Ed.). (1969). *On fighting poverty.* New York: Basic Books.

Sundquist, J. L., & Davis, D. W. (1969). *Making federalism work.* Washington, DC: Brookings Institution.

Susskind, L., McMahon, G., John, E., & Rolley, S. (1986). *Reframing the rationale for downtown linkage policies.* Cambridge, MA: Lincoln Institute of Land Policies.

Tabb, W. K. (1978). The New York city fiscal crisis. In W. K. Tabb & L. Sawers (Eds.), *Marxism and the metropolis: New perspectives in urban political economy* (pp. 241-266). New York: Oxford University Press.

Tabb, W. K., & Sawers, L. (1984). *Marxism and the metropolis: New perspectives in urban political economy* (2nd ed.). New York: Oxford University Press.

Thompson, W. (1977). Land management strategies for central city depopulation. In U.S. House of Representatives (95th Congress, 1st session), *How cities can grow old gracefully: Documents for the house committee on banking, finance, and urban affairs* (pp. 67-78). Washington, DC: Government Printing Office.

Turk, H. (1977). *Organizations in modern life: Cities and other large networks.* San Francisco: Jossey-Bass.

U.S. Bureau of the Census. (1957). *Historical statistics of the United States: Colonial times to 1957.* Washington, DC: Government Printing Office.

U.S. Bureau of the Census. (1981). *Statistical abstract of the United States.* Washington, DC: Government Printing Office.

U.S. Census of Population. (1970). *Number of inhabitants, United States summary, final report.* Washington, DC: Government Printing Office.

U.S. Conference of Mayors. (1993). *Urban policy briefing paper.* Washington, DC: Author.

U.S. Congress. (1974). *1974 Housing and Community Development Act.* Public Law 93-383, § 101, 88 Stat. § 633.

U.S. Department of Housing and Urban Development (U.S. DHUD). (1966). *Improving the quality of urban life: A program guide to model neighborhoods in demonstration cities.* Washington, DC: Government Printing Office.

U.S. Department of Housing and Urban Development (U.S. DHUD). (1969). *Administrative performance and capability* (CDA letter no. 10A). Washington, DC: Government Printing Office.

U.S. Department of Housing and Urban Development (U.S. DHUD). (1970). *Joint HUD-OEO citizen participation policy for model cities programs* (CDA letter no. 10B). Washington, DC: Government Printing Office.

U.S. Department of Housing and Urban Development (U.S. DHUD). (1973). *The model cities program: A comparative analysis of city response patterns.* Washington, DC: Government Printing Office.

U.S. Department of Housing and Urban Development (U.S. DHUD). (1974). *Housing in the seventies.* Washington, DC: Government Printing Office.

U.S. Department of Housing and Urban Development (U.S. DHUD). (1982). *The president's national urban policy report* (HUD-S-702-2). Washington, DC: Government Printing Office.

U.S. Department of Housing and Urban Development (U.S. DHUD). (1992). *Annual report to congress on the community development block grant program.* Washington, DC: Government Printing Office.

U.S. General Accounting Office. (1975). *Fundamental changes are needed in federal assistance to state and local governments.* Washington, DC: Government Printing Office.

U.S. Industrial Commission. (1900-1902). *Reports.* Washington, DC: Government Printing Office.

U.S. Senate. (1982). *S. 2298: The enterprise zone tax act of 1982.* Washington, DC: Author.

U.S. Surplus Property Administration. (1945). *The liquidation of war surplusses.* Washington, DC: Government Printing Office.

Vanecko, J. (1969). Community mobilization and institutional change. *Social Science Quarterly, 50,* 609-630.

Walker, R. W. (1978). The transformation of urban structure in the nineteenth century. In K. R. Cox (Ed.), *Urbanization and conflict in market societies* (pp. 165-212). Chicago: Maaroufa.

Ward, D. N. (1971). *Cities and immigrants: A geography of change in 19th century America.* New York: Oxford University Press.

Warren, C. R. (Ed.). (1985). *Urban policy in a changing federal system.* Washington, DC: National Academy Press.

Warren, R. L. (1967). The interorganizational field as a focus for investigation. *Administrative Science Quarterly, 12,* 396-417.

Warren, R. L. (1977). The sociology of knowledge and the problems of the inner cities. In R. W. Warren (Ed.), *New perspectives on the American community* (pp. 481-500). Chicago: Rand McNally.

Warren, R. L. (1978). *The community in America* (3rd ed.). Chicago: Rand McNally.

Warren, R. L., Rose, S. M., & Bergunder, A. F. (1974). *The structure of urban reform.* Lexington, MA: D. C. Heath.

Wilhelm, S. (1973). The concept of the "ecological complex": A critique. In S. Halebsky (Ed.), *The sociology of the city* (pp. 137-144). New York: Scribner.

Wilson, J. Q. (1967). Planning and politics: Citizen participation in urban renewal. In J. Bellush & M. Hausknecht (Eds.), *Urban renewal: People, politics and planning* (pp. 287-301). Garden City, NY: Doubleday.

Wilson, W. J. (1987). *The truly disadvantaged.* Chicago: University of Chicago Press.

Wirt, F. M. (1973). Alioto and the politics of hyperpluralism. In W. D. Burnham (Ed.), *Politics/America: The cutting edge of change* (pp. 353-362). New York: Van Nostrand.

Wofford, J. G. (1969). The politics of local responsibility: Administration of the community action program, 1964-1966. In J. L. Sundquist (Ed.), *On fighting poverty* (pp. 70-102). New York: Basic Books.

Wood, R. C., & Klimkowsky, B. (1985). Cities in the new federalism. In C. R. Warren (Ed.), *Urban policy in a changing federal system* (pp. 228-253). Washington, DC: National Academy Press.

Woodruff, C. R. (1911). *City government by commission.* New York: Appleton.

Yates, D. (1979). The mayor's eight ring circus: The shape of urban politics in its evolving policy arenas. In D. R. Marshall (Ed.), *Urban policy making* (pp. 41-69). Beverly Hills, CA: Sage.

Zorbaugh, H. (1961). The natural areas of the city. In G. A. Theodorson (Ed.), *Studies in human ecology* (pp. 45-49). New York: Harper & Row.

Index

About the Author

Benjamin Kleinberg is Associate Professor of Sociology at the University of Maryland, Baltimore County (UMBC), where he has taught courses in urban sociology, community organizational systems, and political sociology and has served as chair of the Sociology Department. He also served as acting director of UMBC's multidisciplinary graduate program in policy sciences during a critical period in its establishment. He played a central role in the design and development of the program, developing the curriculum of its sociology track and serving as representative of the sociology department in program governance, as well as teaching core program courses in public policy process. He recently introduced a graduate course in sociological perspectives on public policy and urban development in the program's sociology track and plans to pursue further issues of urban community development as a research interest.

He is the author of *American Society in the Postindustrial Age: Technocracy, Power, and the End of Ideology.* He has worked as a community organizational analyst and field representative doing local area studies and community education for the community renewal section of the New York City Planning Commission and with community groups in New York and Baltimore, including Baltimore Neighborhoods, Inc., a tenant-rights and housing integration research and advocacy organization, and the Baltimore Development Commission, a citizens' advisory group to the mayor on issues relating to community development.